Lecture Notes in Physics

W0112284

The Lecture Notes in Physics

The series Lecture Notes in Physics (LNP), founded in 1969, reports new developments in physics research and teaching – quickly and informally, but with a high quality and the explicit aim to summarize and communicate current knowledge in an accessible way. Books published in this series are conceived as bridging material between advanced graduate textbooks and the forefront of research and to serve three purposes:

- to be a compact and modern up-to-date source of reference on a well-defined topic

- to serve as an accessible introduction to the field to postgraduate students and nonspecialist researchers from related areas

- to be a source of advanced teaching material for specialized seminars, courses and schools

Both monographs and multi-author volumes will be considered for publication. Edited volumes should, however, consist of a very limited number of contributions only. Proceedings will not be considered for LNP.

Volumes published in LNP are disseminated both in print and in electronic formats, the electronic archive being available at springerlink.com. The series content is indexed, abstracted and referenced by many abstracting and information services, bibliographic networks, subscription agencies, library networks, and consortia.

Proposals should be sent to a member of the Editorial Board, or directly to the managing editor at Springer:

Christian Caron
Springer Heidelberg
Physics Editorial Department I
Tiergartenstrasse 17
69121 Heidelberg / Germany
christian.caron@springer.com

Jan-Bert Flór (Ed.)

Fronts, Waves and Vortices in Geophysical Flows

 Springer

Jan-Bert Flor
LEGI (Laboratoire des Ecoulements
Geophysiques et Industriels)
Universite de Grenoble
B.P.53X, 38041 Grenoble Cedex 09
France

Jan-Bert Flor (Ed.): *Fronts, Waves and Vortices in Geophysical Flows*, Lect. Notes Phys.
805 (Springer, Berlin Heidelberg 2010), DOI 10.1007/978-3-642-11587-5

ISBN 978-3-642-11586-8   ISBN 978-3-642-11587-5 (e-Book)
DOI 10.1007/978-3-642-11587-5
Springer Heidelberg Dordrecht London New York

Library of Congress Control Number: 2010922993

over design: Integra Software Services Pvt. Ltd., Pondicherry

Printed on acid-free paper

Springer is part of Springer Science Business Media (www.springer.com)

Foreword

Without coherent structures atmospheres and oceans would be chaotic and unpredictable on all scales of time. Most well-known structures in planetary atmospheres and the Earth oceans are jets or fronts and vortices that are interacting with each other on a range of scales. The transition from one state to another such as in unbalanced or adjustment flows involves the generation of waves, as well as the interaction of coherent structures with these waves. This book presents from a fluid mechanics perspective the dynamics of fronts, vortices, and their interaction with waves in geophysical flows.

It provides a basic physical background for modeling coherent structures in a geophysical context and gives essential information on advanced topics such as spontaneous wave emission and wave-momentum transfer in geophysical flows. The book is targeted at graduate students, researchers, and engineers in geophysics and environmental fluid mechanics who are interested or working in these domains of research and is based on lectures given at the Alpine summer school entitled 'Fronts, Waves and Vortices.' Each chapter is self-consistent and gives an extensive list of relevant literature for further reading. Below the contents of the five chapters are briefly outlined.

Chapter comprises basic theory on the dynamics of vortices in rotating and stratified fluids, illustrated with illuminating laboratory experiments. The different vortex structures and their properties, the effects of Ekman spin-down, and topography on vortex motion are considered. Also, the breakup of monopolar vortices into multiple vortices as well as vortex advection properties will be discussed in conjunction with laboratory visualizations.

In Chap. 2, the understanding of the different vortex instabilities in rotating, stratified, and – in the limit – homogenous fluids are considered in conjunction with laboratory visualizations. These include the shear, centrifugal, elliptical, hyperbolic, and zigzag instabilities. For each instability the responsible physical mechanisms are considered.

In Chap. 3, oceanic vortices as known from various in situ observations and measurements introduce the reader to applications as well as outstanding questions and their relevance to geophysical flows. Modeling results on vortices highlight physical aspects of these geophysical structures. The dynamics of ocean deep sea vortex lenses and surface vortices are considered in relation to their generation mechanism. Further, vortex decay and propagation, interactions as well as the relevance of these processes to ocean processes are discussed. Different types of model equations and the related quasi-geostrophic and shallow water modeling are presented.

In Chap. 4 geostrophic adjustment in geophysical flows and related problems are considered. In a hierarchy of shallow water models the problem of separation of fast and slow variables is addressed. It is shown how the separation appears at small Rossby numbers and how various instabilities and Lighthill radiation break the separation at increasing Rossby numbers. Topics such as trapped modes and symmetric instability, 'catastrophic' geostrophic adjustment, and frontogenesis are presented.

In Chap. 5, nonlinear wave–vortex interactions are presented, with an emphasis on the two-way interactions between coherent wave trains and large-scale vortices. Both dissipative and non-dissipative interactions are described from a unified perspective based on a conservation law for wave pseudo-momentum and vortex impulse. Examples include the generation of vortices by breaking waves on a beach and the refraction of dispersive internal waves by three-dimensional mean flows in the atmosphere.

Grenoble, France   Jan-Bert Flór

Contents

Chapter 1
Dynamics of Vortices in Rotating and Stratified Fluids

G.J.F. van Heijst

The planetary background rotation and density stratification play an essential role in the dynamics of most large-scale geophysical vortices. In this chapter we will discuss some basic dynamical aspects of rotation and stratification, while focusing on elementary vortex structures. Rotation effects will be discussed in Sect. 1.1, attention being given to basic balances, Ekman-layer effects, topography and β-plane effects, and vortex instability. Some laboratory experiments will be discussed in order to illustrate the theoretical issues. Section 1.2 is devoted to vortex structures in stratified fluids, with focus on theoretical models describing their decay. Again, laboratory experiments will play a central part in the discussion. Finally, some general conclusions will be drawn in Sect. 1.3. For additional aspects of the laboratory modelling of geophysical vortices the interested reader is referred to the review papers [14] and [16].

1.1 Vortices in Rotating Fluids

Background rotation tends to make flows two-dimensional, at least when the rotation is strong enough. In this chapter we will discuss some of the basic dynamics of rotating flows and in particular of vortex structures in such flows. After having introduced the basic equations, the principal basic balances will be discussed, followed by some remarks on Ekman boundary layers. Basic knowledge of these topics is important for a better understanding of vortex structures as observed in experiments with rotating fluids, in particular regarding their decay. Further items that will be discussed are topography effects, vortex instability, and advection properties of vortices.

G.J.F. van Heijst (✉)

Deptartment of Applied Physics, Eindhoven University of Technology, PO Box 513, 5600 MB Eindhoven, The Netherlands, g.j.f.v.heijst@tue.nl

van Heijst, G.J.F.: *Dynamics of Vortices in Rotating and Stratified Fluids*. Lect. Notes Phys. **805**, 1–34 (2010)

DOI 10.1007/978-3-642-11587-5_1

1.1.1 Basic Equations and Balances

Flows in a rotating system can be conveniently described relative to a co-rotating reference frame. The position and velocity of a fluid parcel in an inertial frame are denoted by $\mathbf{r}' = (x', y', z')$ and $\mathbf{v}' = \mathbf{v}'(\mathbf{r}')$, respectively, with the primes referring to this particular frame and (x', y', z') being the parcel's coordinates in a Cartesian frame. Relative to a frame rotating about the z'-axis, the position and velocity vectors are $\mathbf{r} = (z, y, z)$ and $\mathbf{v} = \mathbf{v}(\mathbf{r})$, respectively.

For the *velocity* in the inertial frame we write

$$\frac{d\mathbf{r}'}{dt} = \frac{d\mathbf{r}}{dt} + \mathbf{\Omega} \times \mathbf{v} \quad \rightarrow \quad \mathbf{v}' = \mathbf{v} + \mathbf{\Omega} \times \mathbf{r} \tag{1.1}$$

and for the acceleration

$$\frac{d^2\mathbf{r}'}{dt^2} = \frac{d^2\mathbf{r}}{dt^2} + 2\mathbf{\Omega} \times \frac{d\mathbf{r}}{dt} + \mathbf{\Omega} \times \mathbf{\Omega} \times \mathbf{r} \tag{1.2}$$

with

$$2\mathbf{\Omega} \times \frac{d\mathbf{r}}{dt} = 2\mathbf{\Omega} \times \mathbf{v} \qquad \text{Coriolis acceleration} \tag{1.3}$$

$$\mathbf{\Omega} \times \mathbf{\Omega} \times \mathbf{r} = -\nabla \left(\tfrac{1}{2}\Omega^2 r^2 \right) \quad \text{centrifugal acceleration ,} \tag{1.4}$$

where r is the radial distance from the rotation axis, see Fig. 1.1. The equation of motion in terms of the *relative* velocity $\mathbf{v}$ can then be written as

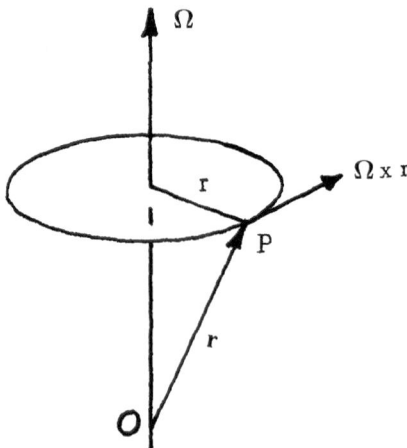

Fig. 1.1 Definition sketch for relative motion in a co-rotating reference frame

$$\frac{D\mathbf{v}}{Dt} + 2\mathbf{\Omega} \times \mathbf{v} = -\frac{1}{\rho}\nabla p - \nabla\Phi + \nu\nabla^2\mathbf{v}, \tag{1.5}$$

with p the pressure, ρ the density, ν the kinematic viscosity, t the time, and

$$\Phi \equiv \Phi_{\mathrm{gr}} - \tfrac{1}{2}\Omega^2 r^2, \tag{1.6}$$

with Φ_{gr} the gravitational potential. By introducing the 'reduced' pressure $P = p - p_{\mathrm{stat}}$, with $p_{\mathrm{stat}} = -\rho\Phi_{\mathrm{gr}} + \tfrac{1}{2}\rho\Omega^2 r^2$, (1.5) can be written as

$$\frac{\partial\mathbf{v}}{\partial t} + (\mathbf{v}\cdot\nabla)\mathbf{v} + 2\mathbf{\Omega}\times\mathbf{v} = -\frac{1}{\rho}\nabla P + \nu\nabla^2\mathbf{v}. \tag{1.7}$$

Together with the continuity equation $\nabla\cdot\mathbf{v} = 0$ for incompressible fluid, this forms the basic equation for rotating fluid flow.

By introducing a characteristic length scale L and a characteristic velocity scale U, the physical quantities are non-dimensionalized according to

$$\mathbf{v} = U\tilde{\mathbf{v}}, \quad \mathbf{r} = L\tilde{\mathbf{r}}, \quad t = \tilde{t}/\Omega, \quad P = \rho\Omega U L\tilde{p}$$

(the tilde indicates the non-dimensional quantity). The non-dimensional form of (1.7) then becomes

$$\frac{\partial\tilde{\mathbf{v}}}{\partial\tilde{t}} + Ro\left(\tilde{\mathbf{v}}\cdot\tilde{\nabla}\right)\tilde{\mathbf{v}} + 2\mathbf{k}\times\tilde{\mathbf{v}} = -\tilde{\nabla}\tilde{p} + E\tilde{\nabla}^2\tilde{\mathbf{v}}, \tag{1.8}$$

with $\mathbf{k} \equiv \mathbf{\Omega}/|\mathbf{\Omega}|$, $\tilde{\nabla}$ the non-dimensional gradient operator, and

$$Ro = \frac{U}{\Omega L} \qquad \text{Rossby number} \tag{1.9}$$

$$E = \frac{\nu}{\Omega L^2} \qquad \text{Ekman number.} \tag{1.10}$$

These non-dimensional numbers provide information about the relative importance of the non-linear advection term and the viscous term, respectively, with respect to the Coriolis term $2\mathbf{\Omega}\times\mathbf{v}$. In the following, we will drop the tildes for convenience.

1.1.1.1 Geostrophic Flow

In many geophysical flow situations both the Rossby number and the Ekman number have very small values, i.e. $Ro \ll 1$ and $E \ll 1$. In the case of steady flow, (1.8) then becomes

$$2\mathbf{k}\times\mathbf{v} = -\nabla p. \tag{1.11}$$

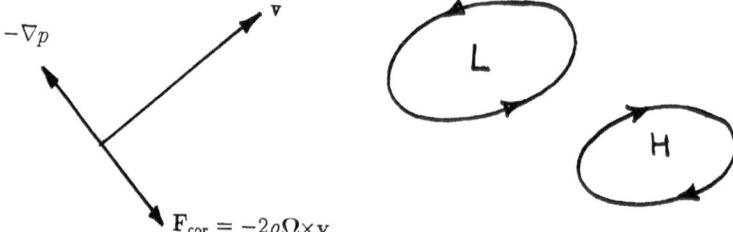

Fig. 1.2 Geostrophically balanced flow on the northern hemisphere

This equation describes flow that is in *geostrophic* balance: the Coriolis force is balanced by the pressure gradient force $(-\nabla p)$. Note that – in dimensional form – the Coriolis force is equal to $-2\rho \boldsymbol{\Omega} \times \mathbf{v}$ and thus acts perpendicular to $\mathbf{v}$, i.e. to the right with respect to a moving fluid parcel (on the northern hemisphere). Apparently, geostrophic motion follows isobars, see Fig. 1.2. For large-scale flows in the atmosphere, $U \simeq 10\,\mathrm{ms}^{-1}$, $L \simeq 1000\,\mathrm{km}$, and $\Omega \simeq 10^{-4}\,\mathrm{s}^{-1}$, which gives $Ro \sim 0.1$. Large-scale oceanic flows are characterized by similarly small Ro values, so that inertial effects are negligibly small in these flows. Likewise, it may be shown that the Ekman numbers of these flows take even smaller values.

By taking the curl of (1.11), we derive

$$(\mathbf{k} \cdot \nabla)\mathbf{v} = 0 \quad \rightarrow \quad \frac{\partial \mathbf{v}}{\partial z} = 0 \,, \tag{1.12}$$

which is the celebrated *Taylor–Proudman theorem*. Apparently, geostrophic motion is independent of the axial coordinate z. Taylor verified this TP theorem (derived by Proudman in 1916) experimentally in 1923 by moving a solid obstacle slowly through a fluid otherwise rotating as a whole. A column of stagnant fluid was observed to be attached to the moving obstacle. This phenomenon is usually referred to as a 'Taylor column'. According to the TP theorem, small Ro flows of a rotating fluid are usually organized in axially aligned columns, i.e. they are uniform in the axial direction.

In most geophysical flow situations, the situation is somewhat more complicated, e.g. by the presence of vertical variations in the density, $\rho(z)$. In each horizontal plane the flow may still be in geostrophic balance (1.11), but because of $\partial \rho / \partial z \neq 0$ the flow is sheared in the vertical. Such a balance is usually referred to as the 'thermal wind balance'.

1.1.1.2 Motion on a Rotating Sphere

The relative flow in the Earth's atmosphere and oceans is most conveniently described when using a local Cartesian coordinate system (x, y, z) fixed to the Earth, with x, y, and z pointing eastwards, northwards, and vertically upwards, respectively. The velocity vector has corresponding components u, v, and w, while the rotation vector can be decomposed as

$$\boldsymbol{\Omega} = (\Omega_x, \Omega_y, \Omega_z) = (0, \Omega \cos \varphi, \Omega \sin \varphi) \,, \tag{1.13}$$

with φ the geographical latitude. Apparently, the term $2\boldsymbol{\Omega} \times \mathbf{v}$ (proportional to the Coriolis acceleration) is then written as

$$2\boldsymbol{\Omega} \times \mathbf{v} = \begin{vmatrix} \mathbf{i} & \mathbf{j} & \mathbf{k} \\ 0 & 2\Omega\cos\varphi & 2\Omega\sin\varphi \\ u & v & w \end{vmatrix} = 2\Omega \begin{pmatrix} w\cos\varphi - v\sin\varphi \\ u\sin\varphi \\ -u\cos\varphi \end{pmatrix} . \qquad (1.14)$$

In the 'thin-shell' approach it is usually assumed that $w << u, v$ for large-scale flows, so that (1.14) becomes

$$2\boldsymbol{\Omega} \times \mathbf{v} = (-fv,\ fu,\ -2\Omega u\cos\varphi) , \qquad (1.15)$$

with $f \equiv 2\Omega\sin\varphi$ the so-called *Coriolis parameter*. It expresses the fact that the background vorticity component in the local z-direction (so perpendicular to the plane-of-flow) varies with latitude φ, being zero on the equator and reaching extreme values at the poles. This directly implies that the magnitude of the Coriolis force also depends on the position (φ) on the rotating globe. The geostrophic balance (1.11) can thus be written (in dimensional form) as

$$- fv = -\frac{1}{\rho}\frac{\partial p}{\partial x} , \quad + fu = -\frac{1}{\rho}\frac{\partial p}{\partial y} . \qquad (1.16)$$

The Coriolis parameter $f(\varphi)$ may be expanded in a Taylor series around the reference latitude φ_0 (see Fig. 1.3):

$$\begin{aligned} f(\varphi) = f(\varphi_0 + \delta\varphi) &= \\ &= 2\Omega \left[\sin\varphi_0 + \frac{\cos\varphi_0}{R} R\delta\varphi + O(\delta\varphi^2) \right] = \\ &= 2\Omega\sin\varphi_0 + \frac{2\Omega\cos\varphi_0}{R} y + \cdots , \end{aligned} \qquad (1.17)$$

with $y = R\delta\varphi$ the local northward coordinate. For flows with limited latitudinal extension, $f(\varphi)$ may be approximated by taking just the first term of the expansion:

$$f = f_0 = 2\Omega\sin\varphi_0 , \qquad (1.18)$$

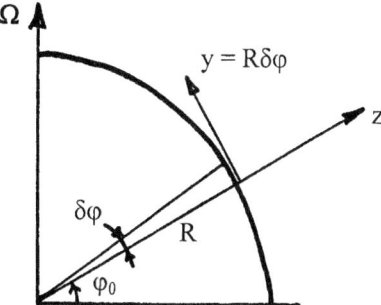

Fig. 1.3 Definition sketch for the expansion of $f(\varphi)$

which is constant. This is the so-called f-*plane* approximation. For flows with larger latitudinal extensions, the Coriolis parameter may be approximated by

$$f = f_0 + \beta y , \quad \beta = \frac{2\Omega \cos \varphi_0}{R} . \tag{1.19}$$

This linear approximation is commonly referred to as the 'beta-plane'.
As will be shown later in this chapter, the latitudinal variation in the Coriolis acceleration has a number of remarkable consequences.

1.1.1.3 Basic Balances

By definition, vortex flows have curvature. In order to examine possible curvature effects we consider a steady, axisymmetric vortex motion in the horizontal plane (assuming that the vortex is columnar). For pure swirling flow the radial and azimuthal velocity components are

$$v_r = 0 , \quad v_\theta = V(r) . \tag{1.20}$$

Following Holton [15] the motion of a fluid parcel along a curved trajectory can be conveniently described in terms of the natural coordinates $\mathbf{n}$ and $\mathbf{t}$ in the local normal and tangential directions and by defining the local radius of curvature, R (see Fig. 1.4). Keeping in mind that $R > 0$ relates to anti-clockwise motion (cyclonic, on the NH), whereas $R < 0$ refers to clockwise motion. For steady inviscid flow with circular streamlines, the equation of motion (in dimensional form) is then simply

$$\frac{V^2}{R} + fV = -\frac{1}{\rho}\frac{dp}{dn} . \tag{1.21}$$

This equation represents a balance between centrifugal, Coriolis, and pressure gradient forces. In non-dimensional form, the Rossby number would appear in front of the centrifugal acceleration term V^2/R. We will now examine the effect of this

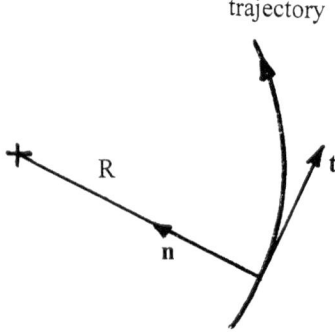

Fig. 1.4 Definition sketch for the natural coordinates $\mathbf{n}$ and $\mathbf{t}$

curvature term by varying the value of the Rossby number

$$Ro^* = \frac{[(\mathbf{v} \cdot \nabla)\mathbf{v}]}{[2\boldsymbol{\Omega} \times \mathbf{v}]} = \frac{V^2/R}{fV} = \frac{V}{\Omega R}, \tag{1.22}$$

which is in fact a local Rossby number.

(i) $Ro^* << 1$: *geostrophic balance*
Equation (1.21) reduces to

$$fV = -\frac{1}{\rho}\frac{dp}{dn}, \tag{1.23}$$

which is the well-known geostrophic balance . For $\frac{dp}{dn} < 0$ it describes the cyclonic motion around a centre of low pressure, while $\frac{dp}{dn} > 0$ corresponds with anticyclonic flow around a high-pressure area.

(ii) $Ro^* >> 1$: *cyclostrophic balance*
In this case the Coriolis term is negligibly small (compared to the centrifugal term) and (1.21) becomes

$$\frac{V^2}{R} = -\frac{1}{\rho}\frac{dp}{dn} \quad \rightarrow \quad V = \pm\left(-\frac{R}{\rho}\frac{dp}{dn}\right)^{1/2}. \tag{1.24}$$

Apparently this balance only exists for the case $\frac{dp}{dn} < 0$, with the outward centrifugal force being balanced by the inward pressure gradient force. The rotation can be in either direction (the sign of V is irrelevant in the term V^2/R). This balance is encountered, e.g. in an atmospheric tornado, with typical values of $V \simeq 30 \text{ ms}^{-1}$ at a radius $R \simeq 300$ m and $f \simeq 10^{-1} \text{ s}^{-1}$ (at moderate latitude) giving $Ro^* \simeq 10^3$.
Similarly large Ro^* values are met in a bathtub vortex, whose rotation sense is obviously *not* determined by the Earth rotation.

(iii) $Ro^* = O(1)$: *gradient flow*
In this case all terms in (1.21) are equally important, and the solution for V is

$$v = -\frac{1}{2}fR \pm \left[\frac{1}{4}f^2R^2 - \frac{R}{\rho}\frac{dp}{dn}\right]^{1/2}. \tag{1.25}$$

This solution represents four different balances, which are shown schematically in Fig. 1.5. Only the flows depicted in (a) and (b) are 'regular', the other two being 'anomalous'.

Note that in order to have a non-imaginary solution, the pressure gradient is required to have a value

$$\left|\frac{dp}{dn}\right| < \frac{1}{2}\rho|R|f^2. \tag{1.26}$$

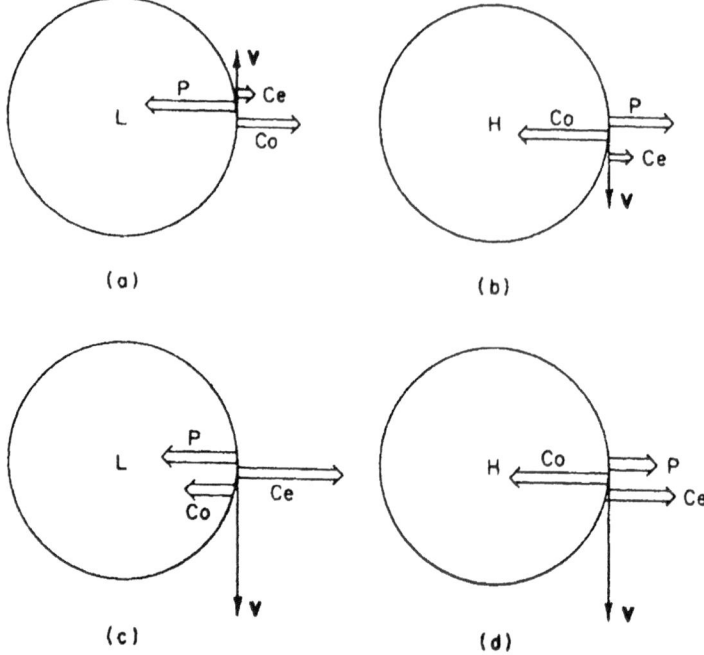

Fig. 1.5 Different balances in gradient flow on the NH: (**a**) regular low, (**b**) regular high, (**c**) anomalous low, and (**d**) anomalous high [after Holton, 1979]

1.1.1.4 Inertial Motion

A special balanced state may exist in the absence of any pressure gradient, i.e. when $\frac{dp}{dn} = 0$. In that case (1.21) becomes

$$\frac{V^2}{R} + fV = 0 \,, \tag{1.27}$$

which describes so-called inertial motion. Fluid parcels move with constant speed V (the solution $V = 0$ is trivial and physically uninteresting) along a circular path with radius $R = -V/f < 0$, i.e. in *anticyclonic* direction. The centrifugal force is then exactly balanced by the inward Coriolis force. In x, y-coordinates, the motion can be described by

$$u(t) = V \cos ft \,, \quad v(t) = -V \sin ft \,, \quad \text{with} \ \ V = (u^2 + v^2)^{1/2} \,.$$

The time required for the fluid parcels to perform one circular orbit is the so-called *inertial period*, which is equal to $T = 2\pi/f$.

1.1.2 How to Create Vortices in the Lab

A barotropic vortex can be generated in a rotating fluid in a number of different ways. One possible way is to place a thin-walled bottomless cylinder in the rotating fluid and then stir the fluid inside this cylinder, either cyclonically or anticyclonically. After allowing irregular small-scale motions to vanish and the vortex motion to get established (which typically takes a few rotation periods) the vortex is released by quickly lifting the cylinder out of the fluid. The vortex structure thus created in the otherwise rigidly rotating fluid is referred to as a 'stirring vortex'. Because these vortices are generated within a solid cylinder with a no-slip wall, the total circulation – and hence the total vorticity – measured in the rotating frame is zero, i.e. stirring vortices are *isolated* vortex structures:

$$\Gamma = \oint_c \mathbf{v} \cdot d\mathbf{r} = \int\int_A \omega_z dA = 0 . \tag{1.28}$$

An alternative way of generating vortices is to have the fluid level in the inner cylinder lower than outside it (see Fig. 1.6): the 'gravitational collapse' that takes place after lifting the cylinder implies a radial inward motion of the fluid, which by conservation of angular momentum results in a cyclonic swirling motion. After any small-scale and wave-like motions have vanished, the swirling motion takes on the appearance of a columnar vortex. In contrast to the stirring vortices, these 'gravitational collapse vortices' have a non-zero net vorticity and are hence not isolated. This technique as well as the generation technique of stirring vortices has been applied successfully by Kloosterziel and van Heijst [18] in their study of the evolution of barotropic vortices in a rotating fluid.

A related generation method has recently been used by Cariteau and Flór [4]: they placed a solid cylindrical bar in the fluid and after pulling it vertically upwards

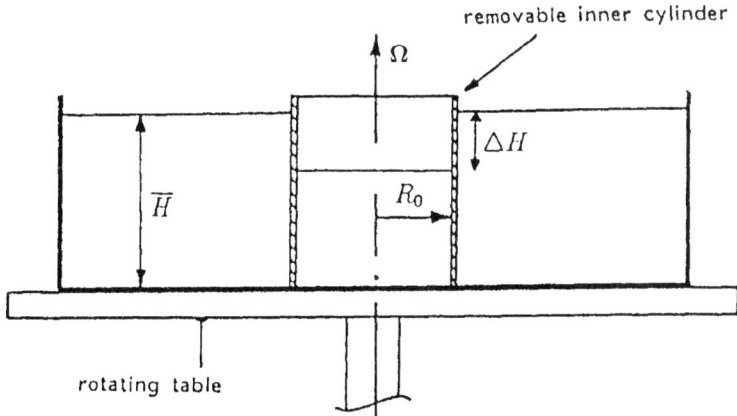

Fig. 1.6 Laboratory arrangement for the creation of barotropic vortices

the resulting radial inward motion of the fluid was quickly converted into a cyclonic swirling flow, as in the previous case.

Another vortex generation technique is based on removing some of the rotating fluid from the tank by syphoning through a vertical, perforated tube. Again, the suction-induced radial motion is quickly converted into a cyclonic swirling motion – owing to the principle of conservation of angular momentum. This generation technique has been applied by Trieling et al. [24], who showed that – outside its core – the 'sink vortex' has the following azimuthal velocity distribution:

$$v_\theta(r) = \frac{\gamma}{2\pi r} \left[1 - \exp\left(-\frac{r^2}{L^2}\right) \right], \tag{1.29}$$

with γ the total circulation of the vortex and L a typical radial length scale. Vortices have also been created in a rotating fluid by translating or rotating vertical flaps through the fluid. Alternatively, buoyancy effects may also lead to vortices in a rotating fluid, as seen, e.g. in experiments with a melting ice cube at the free surface (see, e.g. Whitehead et al. [29] and Cenedese [7]) or by releasing a volume of denser or lighter fluid (see, e.g. Griffiths and Linden [12]).

In all these cases, the vortices are observed to have a columnar structure and $\frac{\partial v_\theta}{\partial z} = 0$, as follows from the TP theorem, even for larger Ro values. Viscosity is responsible for the occurrence of an Ekman layer at the tank bottom, in which the vortex flow is adjusted to the no-slip condition at the solid bottom. Ekman layers play an important role in the spin-down (or spin-up) of vortices. Kloosterziel and van Heijst [18] have studied the decay of barotropic vortices in a rotating fluid in detail. It was found that this type of vortex, as well as the stirring-induced vortex, is characterized by the following radial distributions of vorticity and azimuthal velocity:

$$\omega_{\text{stir}}(r) = \omega_0 \left(1 - \frac{r^2}{R^2} \right) \exp\left(-\frac{r^2}{R^2}\right), \tag{1.30a}$$

$$v_{\text{stir}}(r) = \frac{\omega_0 r}{2} \exp\left(-\frac{r^2}{R^2}\right). \tag{1.30b}$$

The velocity data in Fig. 1.7a–d have been fitted with (1.30b), which shows a very good correspondence.

Similarly, velocity data of decaying sink-induced vortices turned out to be well fitted (see Kloosterziel and van Heijst [18]; Fig. 1.4) by

$$\omega_{\text{sink}}(r) = \omega_0 \exp\left(-\frac{r^2}{R^2}\right), \tag{1.31a}$$

$$v_{\text{sink}}(r) = \frac{\omega_0 R^2}{2r} \left[1 - \exp\left(-\frac{r^2}{R^2}\right) \right]. \tag{1.31b}$$

Note that for large r values ($r \gg R$) this azimuthal velocity distribution agrees with (1.29).

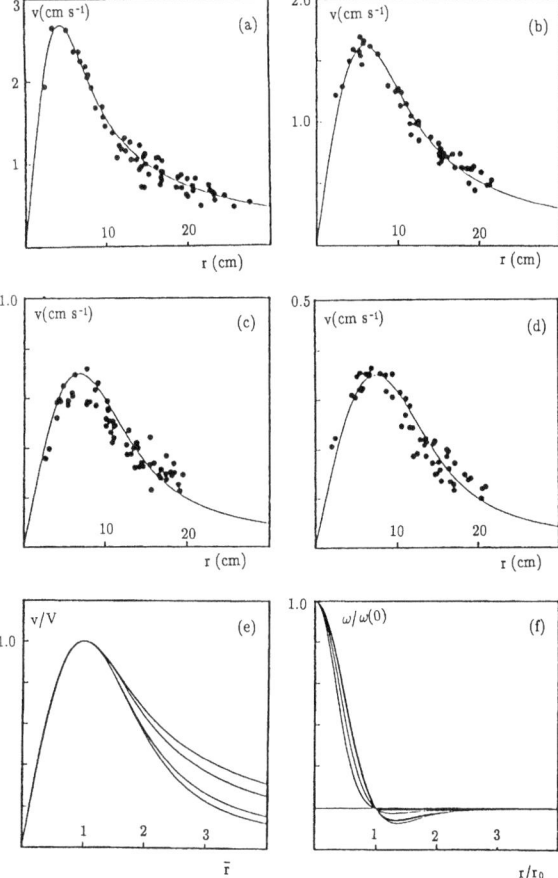

Fig. 1.7 Evolution of collapse-induced vortices in a rotating tank (from [18])

Although vortices with a velocity profile (1.31b) were found to be stable, Carton and McWilliams [6] have shown that those with velocity profile (1.30b) are linearly unstable to $m = 2$ perturbations. It may well be, however, that the instability is not able to develop when the decay (spin-down) associated with the Ekman-layer action is sufficiently fast. In the viscous evolution of stable vortex structures two effects play a simultaneous role: the spin-down due to the Ekman layer, with a timescale

$$T_E = \frac{H}{(\nu\Omega)^{1/2}} \qquad (1.32)$$

and the diffusion of vorticity in radial direction, which takes place on a timescale

$$T_d = \frac{L^2}{\nu}, \qquad (1.33)$$

with H the fluid depth and L a measure of the core size of the vortex. For typical values $\nu = 10^{-6}$ m^{2}s^{-1}, $\Omega \sim 1$ s^{-1}, $L \sim 10^{-1}$ m, and $H = 0.2$ m one finds

$$T_d \sim 10^4 \text{ s}, \quad T_E \sim 2 \cdot 10^2 \text{ s}. \tag{1.34}$$

Apparently, in these laboratory conditions the effects of radial diffusion take place on a very long timescale and can hence be neglected. For a more extensive discussion of the viscous evolution of barotropic vortices, the reader is referred to [18] and [20].

1.1.3 The Ekman Layer

For steady, small-Ro flow (1.8) reduces to

$$2\mathbf{k} \times \mathbf{v} = -\nabla p + E\nabla^2 \mathbf{v}, \tag{1.35}$$

with the last term representing viscous effects. Although E is very small, this term may become important when large velocity gradients are present somewhere in the flow domain. This is the case, for example, in the Ekman boundary layer at the tank bottom, where

$$E\nabla^2 \sim E\frac{\partial^2}{\partial z^2} \sim O(1). \tag{1.36}$$

Apparently the non-dimensional layer thickness is $\delta_E \sim E^{1/2}$ and hence in dimensional form

$$L\delta_E = LE^{1/2} = \left(\frac{\nu}{\Omega}\right)^{1/2}. \tag{1.37}$$

In a typical rotating tank experiment we have $\nu = 10^{-6}$ m^{2}s^{-1} (water), $\Omega \simeq 1$ s^{-1}, and $L \simeq 0.3$ m, so that $E \sim 10^{-5}$, and hence $LE^{1/2} \sim 10^{-3}$ m $= 1$ mm. The Ekman layer is thus very thin.

Since the (non-dimensional) horizontal velocities in the Ekman layer are $O(1)$, the Ekman layer produces a horizontal volume flux of $O(E^{1/2})$. In the Ekman layer underneath an axisymmetric, columnar vortex, this transport has both an azimuthal and a radial component. Mass conservation implies that the Ekman layer consequently produces an axial $O(E^{1/2})$ transport, depending on the net horizontal convergence/divergence in the layer. According to this mechanism, the Ekman layer imposes a condition on the interior flow. This so-called *suction condition* relates the vertical $O(E^{1/2})$ velocity to the vorticity ω_I of the interior flow:

$$w_E(z = \delta_E) = \frac{1}{2}E^{1/2}(\omega_I - \omega_B) \tag{1.38}$$

with

$$\omega_I = \frac{1}{r}\frac{\partial}{\partial r}(r v_{\theta_I}) - \frac{1}{r}\frac{\partial v_{r_I}}{\partial \theta} \tag{1.39}$$

and ω_B the relative bottom 'vorticity'. For example, in the case of a cyclonic vortex ($\omega_I > 0$) over a tank bottom that is at rest in the rotating frame ($\omega_B = 0$), the suction condition yields $w_E(z = \delta_E) > 0$: this corresponds with a radially inward Ekman flux (cf. Einstein's 'tea leaves experiment'), resulting in Ekman blowing, see Fig. 1.8a. In the case of an anticyclonic vortex, the suction condition gives $w_E = (z = \delta_E) < 0$, see Fig. 1.8b.

In the case of an *isolated vortex*, like the stirring-induced vortex with vorticity profile (1.30a), the Ekman layer produces a rather complicated circulation pattern, with vertical upward motion where $\omega_I > 0$ and vertical downward motion where $\omega_I < 0$. This secondary $O(E^{1/2})$ circulation, although weak, results in a gradual change in the vorticity distribution $\omega_I(r)$ in the vortex.

According to this mechanism, a vortex may gradually change from a stable into an unstable state, as was observed for the case of a cyclonic, stirring-induced barotropic vortex [17]. Although this vortex was initially stable, the Ekman-driven $O(E^{1/2})$ circulation resulted in a gradual steepening of velocity/vorticity profiles so that the vortex became unstable and soon transformed into a tripolar structure.

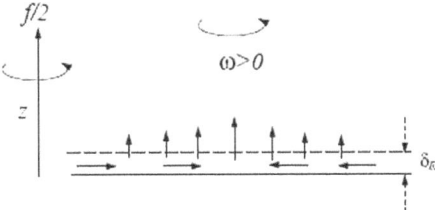

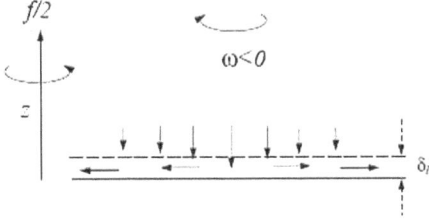

Fig. 1.8 Ekman suction or blowing, depending on the sign of the vorticity of the interior flow

1.1.4 Vortex Instability

Figure 1.9 shows a sequence of photographs illustrating the instability of a cyclonic barotropic isolated vortex as observed in the laboratory experiment by Kloosterziel and van Heijst [17]. In this experiment, the cyclonic stirring-induced vortex was released by vertically lifting the inner cylinder, and although this release process produced some 3D turbulence the vortex soon acquired a regular appearance, as can be seen in the smooth distribution of the dye. Then a shear instability developed with the negative vorticity of the outer edge of the vortex accumulating in two satellite vortices, while the positive-vorticity case acquired an elliptical shape. The newly formed tripolar vortex rotates steadily about its central axis and was observed to be quite robust. This 2D shear instability resulted in a redistribution of the positive and negative vorticities and is very similar to what Flierl [9] found in his stability study of vortex structures with discrete vorticity levels. In a similar experiment, but

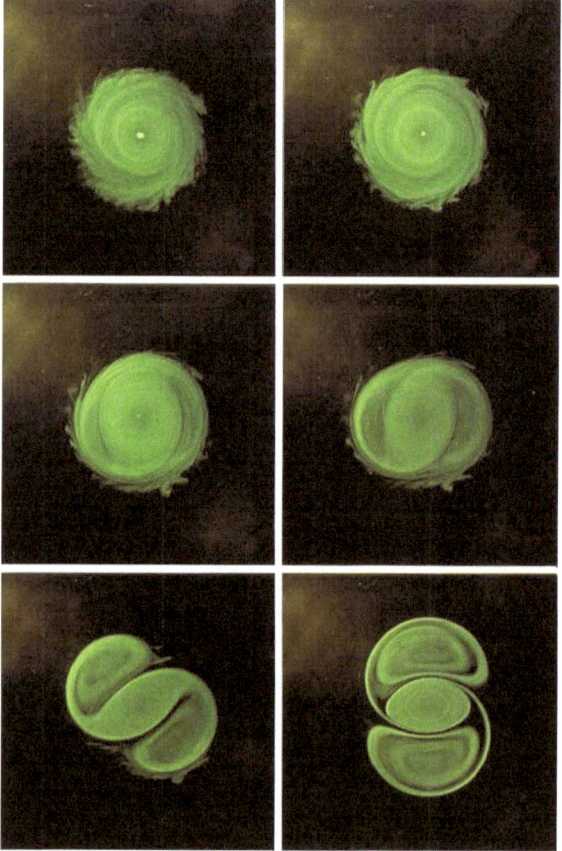

Fig. 1.9 Sequence of photographs illustrating the transformation of an unstable cyclonic vortex (generated with the stirring method) into a tripolar vortex structure (from [17])

now with the stirring in anticyclonic direction, the anticyclonic vortex appeared to be highly unstable, quickly showing vigorous 3D overturning motions (after which two-dimensionality was re-established by the background rotation, upon which the flow became organized in two non-symmetric dipolar vortices, see Fig. 1.5 in [17]). The 3D overturning motions in the initial anticyclonic vortex are the result of a 'centrifugal instability'. Based on energetic arguments, Rayleigh analysed the stability of axisymmetric swirling flows, which led to his celebrated circulation theorem. According to Rayleigh's circulation theorem a swirling flow with azimuthal velocity $v(r)$ is *stable* to axisymmetric disturbances provided that

$$\frac{d}{dr}(rv)^2 \geq 0 .\tag{1.40}$$

This analysis has been extended by Kloosterziel and van Heijst [17] to a swirling motion in an otherwise solidly rotating fluid (angular velocity $\Omega = \frac{1}{2}f$), leading to the following criterion:

$$\frac{d}{dr}(rv + \frac{1}{2}fr)^2 \geq 0 \qquad \text{stable} .\tag{1.41}$$

The modified Rayleigh criterion can also be written as

$$\begin{aligned}(v + \Omega r)(\omega + 2\Omega) &\geq 0 \text{ stable} \\ &< 0 \text{ unstable} ,\end{aligned}\tag{1.42}$$

implying stability if $v_{abs}\omega_{abs} > 0$ at all positions r in the vortex flow. Kloosterziel and van Heijst [17] applied these criteria to the sink-induced and the stirring-induced vortices discussed earlier, with distributions of vorticy and azimuthal velocity given by (1.31a, b) and (1.30a, b), respectively.

It was found that cyclonic *sink*-induced vortices are always stable to axisymmetric disturbances, while their anticyclonic counterparts become unstable for Rossby number values $Ro \gtrsim 0.57$, with the Rossby number $Ro = V/\Omega R$ based on the maximum velocity V and the radius $r = R$ at which this maximum occurs.

For the *stirring*-induced vortices it was found that the cyclonic ones are unstable for $Ro \gtrsim 4.5$ while the anticylonic vortices are unstable for $Ro \gtrsim 0.65$. As a rule of thumb, these results for isolated vortices may be summarized as follows:

- only very weak anticyclonic vortices are centrifugally stable;
- only very strong cyclonic vortices are centrifugally unstable.

1.1.5 Evolution of Stable Barotropic Vortices

Assuming planar motion $\mathbf{v} = (u, v)$, the x, y-components of (1.7) can, after using (1.15), be written as

$$\frac{\partial u}{\partial t} + u\frac{\partial u}{\partial x} + v\frac{\partial u}{\partial y} - fv = -\frac{1}{\rho}\frac{\partial P}{\partial x} + \nu\nabla^2 u \qquad (1.43a)$$

$$\frac{\partial v}{\partial t} + u\frac{\partial v}{\partial x} + v\frac{\partial v}{\partial y} + fu = -\frac{1}{\rho}\frac{\partial P}{\partial y} + \nu\nabla^2 v \ . \qquad (1.43b)$$

By taking the x-derivative of (1.43b) and subtracting the y-derivative of (1.43a) one obtains the following equation for the vorticity $\omega = \frac{\partial v}{\partial x} - \frac{\partial u}{\partial y}$:

$$\frac{\partial \omega}{\partial t} + u\frac{\partial \omega}{\partial x} + v\frac{\partial \omega}{\partial y} + \left(\frac{\partial u}{\partial x} + \frac{\partial v}{\partial y}\right)(\omega + f) = \nu\nabla^2\omega \ . \qquad (1.44)$$

Integration of the continuity equation over the layer depth H yields

$$\int_0^H \nabla \cdot \mathbf{v}dz \ \Rightarrow \ \left(\frac{\partial u}{\partial x} + \frac{\partial v}{\partial y}\right)H = -w(z = H) + w(z = 0) \ . \qquad (1.45)$$

Assuming a flat, non-moving free surface one has $w(z = H) = 0$, while the suction condition (1.38) imposed by the Ekman layer at the bottom yields $w(z = 0) = \frac{1}{2}E^{1/2}\omega$. The vorticity equation (1.44) then takes the following form:

$$\frac{\partial \omega}{\partial t} + u\frac{\partial \omega}{\partial x} + v\frac{\partial \omega}{\partial y} = \nu\nabla^2\omega - \frac{1}{2}E^{1/2}\omega(\omega + f) \ . \qquad (1.46)$$

When the Rossby number $Ro = |\omega|/f$ is small (i.e. for very weak vortices), the nonlinear Ekman condition is usually replaced by its linear version $-\frac{1}{2}E^{1/2}f\omega$. For moderate Ro values, as encountered in most practical cases, however, one should keep the nonlinear condition. A remarkable feature of this nonlinear condition is the symmetry breaking associated with the term $\omega(\omega + f)$: it appears that cyclonic vortices ($\omega > 0$) show a faster decay than anticyclonic vortices ($\omega < 0$) with the same Ro value.

The vorticity equation (1.46) can be further refined by including the weak $O(E^{1/2})$ circulation driven by the bottom Ekman layer, as also schematically indicated in Fig. 1.8. This was done by Zavala Sansón and van Heijst [32], resulting in

$$\frac{\partial \omega}{\partial t} + J(\omega, \Psi) - \frac{1}{2}E^{1/2}\nabla\Psi \cdot \nabla\omega = \nu\nabla^2\omega - \frac{1}{2}E^{1/2}\omega(\omega + f) \ , \qquad (1.47)$$

with J the Jacobian operator and ψ the streamfunction, defined as $\mathbf{v} = \nabla \times (\psi\mathbf{k})$, with $\mathbf{k}$ the unit vector in the direction perpendicular to the plane of flow. These authors have examined the effect of the individual Ekman-related terms in (1.47) by numerically studying the time evolution of a sink-induced vortex for various cases: with and without the $O(E^{1/2})$ advection term, with and without the (non)linear Ekman term. Not surprisingly, the best agreement with experimental observations was obtained with the full version (1.47) of the vorticity equation.

The action of the individual Ekman-related terms in (1.47) can also be nicely examined by studying the evolution of a barotropic dipolar vortex. In the laboratory such a vortex is conveniently generated by dragging a thin-walled bottomless cylinder slowly through the fluid, while gradually lifting it out. It turns out that for slow enough translation speeds the wake behind the cylinder becomes organized in a columnar dipolar vortex. Flow measurements have revealed that this vortex is in very good approximation described by the Lamb–Chaplygin model (see [21]) with the dipolar vorticity structure confined in a circular region, satisfying a linear relationship with the streamfunction, i.e. $\omega = c\psi$. Zavala Sansón et al. [31] have performed

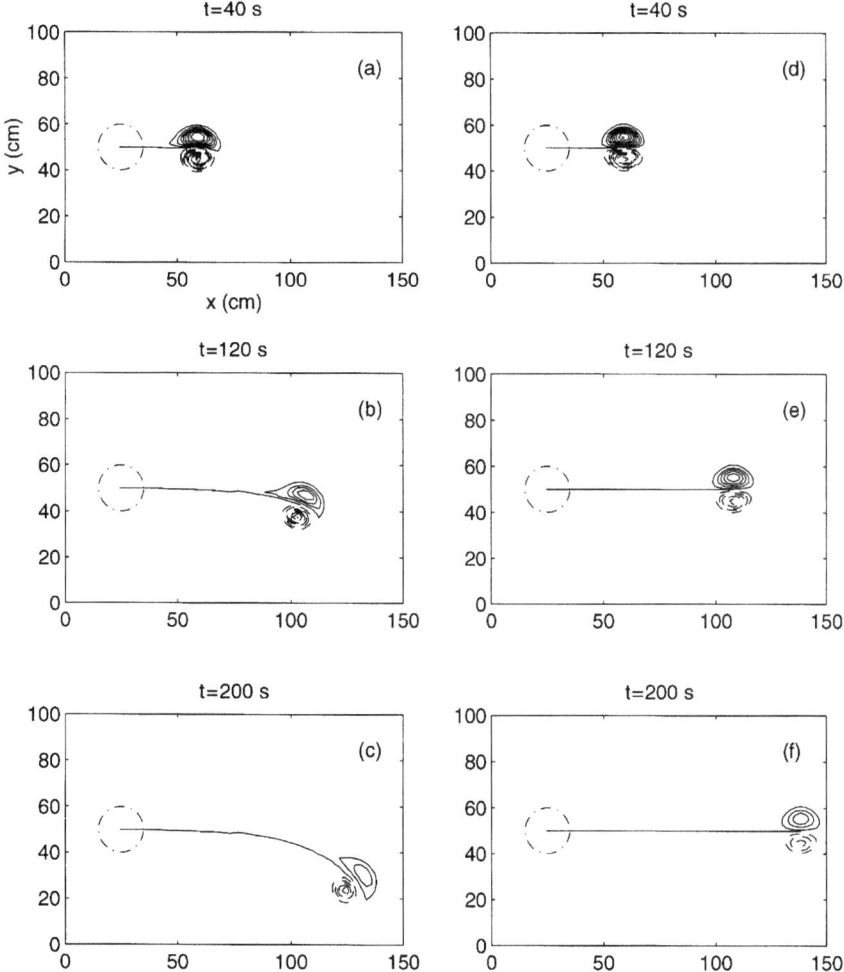

Fig. 1.10 Sequence of vorticity snapshots obtained by numerical simulation of the Lamb–Chaplygin dipole based on (1.46), both for nonlinear Ekman term (*left column*) and linear Ekman term (*right column*). Reproduced from Zavala Sansón et al. [31]

numerical simulations based on the vorticity equation (1.46), both for the linear and
for the nonlinear terms. When the nonlinear term is included, the difference in decay
rates of cyclonic and anticyclonic vortices becomes clearly visible in the increasing
asymmetry of the dipolar structure: its anticyclonic half becomes relatively stronger,
thus resulting in a curved trajectory of the dipole, see Fig. 1.10.

1.1.6 Topography Effects

Consider a vortex column in a layer of fluid that is rotating with angular velocity
Ω. Assuming that viscous effects play a minor role on the timescale of the flow
evolution that we consider here, Helmholtz' theorem applies:

$$\frac{\omega_{abs}}{H} = \frac{2\Omega + \omega}{H} = \text{constant} , \tag{1.48}$$

where ω_{abs} and ω are the absolute and relative vorticities and H the column height
(= fluid depth). This conserved quantity $(2\Omega + \omega)/H$ is commonly referred to as
the potential vorticity. Apparently, a change in the column height H (see Fig. 1.11)
results in a change in the relative vorticity. The term 2Ω in (1.48) implies a symmetry breaking, in the sense that cyclonic and anticyclonic vortices behave differently
above the same topography: a *cyclonic* vortex ($\omega > 0$) moving into a shallower
area becomes *weaker*, while an *anticyclonic* vortex ($\omega < 0$) moving into the same
shallower area becomes *more intense*.

In the so-called shallow-water approximation the large-scale motion in the atmosphere or the ocean can be considered as organized in the form of fluid or vortex
columns that are oriented in the local vertical direction, see Fig. 1.12. For each
individual column the potential vorticity is conserved (as in the case considered
above), taking the following form:

$$\frac{f + \omega}{H} = \text{constant}, \tag{1.49}$$

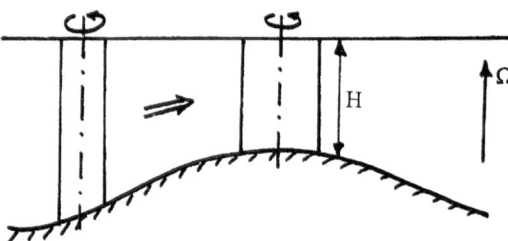

Fig. 1.11 Stretching or squeezing of vortex columns over topography results in changes in the
relative vorticity

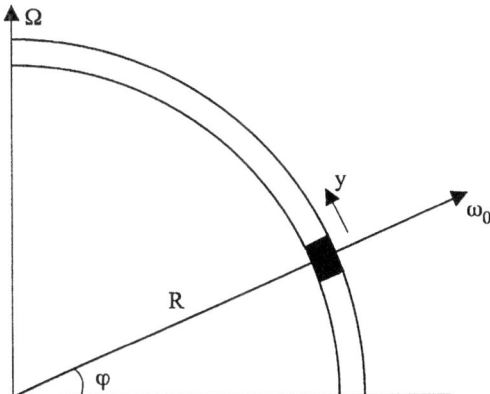

Fig. 1.12 Vortex column in a spherical shell (ocean, atmosphere) covering a rotating sphere

with $f = 2\Omega \sin \varphi$ the Coriolis parameter, as introduced in (1.15), and H the local column height. It should be kept in mind that the vortex columns, and hence the relative-vorticity vector, are oriented in the local vertical direction, so that their absolute vorticity is $(2\Omega \sin \varphi + \omega)$, the first term being the component of the planetary vorticity in the local vertical direction.

In order to demonstrate the implications of conservation of potential vorticity (1.49) on large-scale geophysical flows, we consider a vortex in a fluid layer with a constant depth H_0. When this vortex is shifted northwards, f increases in order to keep $(f + \omega)/H_0$ constant. Here we meet the same asymmetry due to the background vorticity as in the topography case discussed above: a *cyclonic* vortex ($\omega_0 > 0$) moving northwards becomes *weaker*, while an *anticyclonic* vortex ($\omega_0 < 0$) will *intensify* when moving northwards. This is usually referred to as asymmetry caused by the 'β-effect', i.e. the gradient in the planetary vorticity.

Conservation of potential vorticity, as expressed by (1.49), can now be exploited to model the planetary β-effect in a rotating tank by a suitably chosen bottom topography. Changes of the Coriolis parameter f with the northward coordinate y, as in the β-plane approximation $f(y) = f_0 + \beta y$, see (1.19), can be dynamically mimicked in the laboratory by a variation in the water depth $H(y)$, according to

$$\frac{f(y) + \omega}{H_0} = \frac{f_0 + \omega}{H(y)} = \text{constant} ,$$

$$\text{GFD} \qquad \text{LAB}$$

(1.50)

with H_0 the constant fluid depth in the geophysical case (GFD) and $f_0 = 2\Omega$ the constant Coriolis parameter in the rotating tank experiment (LAB). In general, moving into shallower water in the rotating fluid experiment corresponds with moving northwards in the GFD case. It can be shown (see, e.g. [13]) that for small Ro values and weak topography effects (small amplitude: $\Delta h << H$, and weak slopes ∇h)

the β-plane approximation $f(y) = f_0 + \beta y$ can be simply modelled by a uniformly sloping bottom in a rotating fluid tank. This situation is commonly referred to as the 'topographic β-plane'. Since the motion of a fluid column or parcel on a (topographic) β-plane implies changes in its relative vorticity, the following question a rises: How will a vortex structure on a (topographic) β-plane behave? Let us first consider a simple, axisymmetric (monopolar) vortex motion. Obviously, on an f-plane ($f = f_0$) such changes in ω are not introduced and hence the vortex flow is unaffected. The situation on a β-plane is essentially different, however: the relative vorticity ω of fluid parcels in the primary vortex flow that are advected northwards will decrease, while that of southward advected parcels will increase. As a result, a dipolar perturbation will be imposed on the primary vortex, which will result in a *drift* of the vortex structure. This drift has a westward component (i.e. with the 'north' or 'shallow' on its right), the cyclonic vortices drifting in NW direction and the anticyclonic ones moving in SW direction. For a more detailed account on this topographic drift, the reader is referred to Carnevale et al. [5].

The motion of a dipolar vortex on a β-plane is even more intricate. Due to its self-propelling mechanism, a symmetric dipole on an f-plane will move along a straight trajectory. When released on a β-plane, any northward/southward motion of the dipolar structure implies changes in the relative vorticity, i.e. changes in the strengths of the dipole halves: when moving with a northward component the cyclonic part of the dipole will become weaker, while the anticyclonic part intensifies. As a result, the dipole becomes asymmetric and starts to move along a curved trajectory. Depending on the orientation angle at which the dipole is released with respect to the east–west axis, it may perform a meandering motion towards the east or a cycloid-like motion in the western direction. This behaviour, which was confirmed experimentally by Velasco Fuentes and van Heijst [27], may be modelled in a simple way by applying a so-called modulated point-vortex model, in which the strengths of the vortices are made functions of the northward coordinate y. For further details on this type of modelling, the reader is referred to Zabusky and McWilliams [30] and Velasco Fuentes et al. [28].

1.2 Vortices in Stratified Fluids

The dynamics of many large-scale geophysical flows is essentially influenced by density stratification. In this section we will pay some attention to one specific type of flows, viz. the dynamics of pancake-shaped monopolar vortices.

1.2.1 Basic Properties of Stratified Fluids

In order to reveal some basic properties of density stratification we carry out the following 'thought experiment': in a linearly stratified fluid column we displace a little fluid parcel vertically upwards over a distance ζ, see Fig. 1.13. How will this

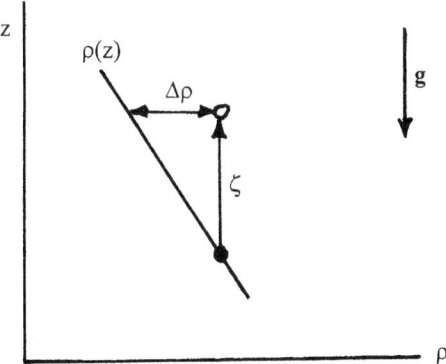

Fig. 1.13 Schematic diagram of the virtual experiment with the displaced fluid parcel

parcel move when released? In this ideal experiment it is assumed that no mixing occurs between the displaced parcel and the ambient. By displacing the parcel over a vertical distance ζ it is introduced in an ambient with a smaller density, the density difference being

$$\delta\rho = -\frac{d\rho}{dz}\zeta \ . \tag{1.51}$$

The downward restoring gravity force (per unit volume) is then

$$g\zeta\frac{d\rho}{dz} \ , \tag{1.52}$$

which – in absence of any other forces – results in a vertical acceleration $\frac{d^2\zeta}{dt^2}$. The equation of motion for the displaced parcel is then

$$\rho\frac{d^2\zeta}{dt^2} = g\frac{d\rho}{dz}\zeta \tag{1.53}$$

or

$$\frac{d^2\zeta}{dt^2} + N^2\zeta = 0, \tag{1.54}$$

with

$$N^2 \equiv -\frac{g}{\rho}\frac{d\rho}{dz} \ . \tag{1.55}$$

The quantity N is usually referred to as the 'buoyancy frequency'. For a statically *stable stratification* ($\frac{d\rho}{dz} < 0$) this frequency N is real, and the solutions of (1.54) take the form of harmonic oscillations. For example, for the initial condition $\zeta(t = 0) = \zeta_0$ and $\zeta'(t = 0) = 0$ the solution is $\zeta(t) = \zeta_0 \cos Nt$, which describes an undamped wave with the natural frequency N. Addition of some viscous damping leads to a damped oscillation, with the displaced parcel finally ending at its original level $\zeta = 0$. Apparently, this stable stratification supports wavelike motion, but vertical mixing is suppressed.

For an *unstable stratification* ($\frac{d\rho}{dz} > 0$) the buoyancy frequency is purely imaginary, i.e. $N = i\overline{N}$, with $\overline{N}$ real. For the same initial conditions the solution of (1.54) now has the following form:

$$\zeta(t) = \frac{1}{2}\zeta_0(e^{-\overline{N}t} + e^{\overline{N}t}) \,. \qquad (1.56)$$

The latter term has an explosive character, representing strong overturning flows and hence mixing. In what follows we concentrate on vortex flows in a stably stratified fluid.

1.2.2 Generation of Vortices

Experimentally, vortices may be generated in a number of different ways, some of which are schematically drawn in Fig. 1.14. Vortices are easily produced by localized stirring with a rotating, bent rod or by using a spinning sphere. In both cases the rotation of the device adds angular momentum to the fluid, which is swept outwards by centrifugal forces. After some time the rotation of the device is stopped, upon which it is lifted carefully out of the fluid. It usually takes a short while for the turbulence introduced during the forcing to decay, until a laminar horizontal vortex motion results. The shadowgraph visualizations shown in Fig. 1.15 clearly reveal the turbulent region during the forcing by the spinning sphere and the more smooth density structure soon after the forcing is stopped. Vortices produced in this way (either with the spinning sphere or with the bent rod) typically have a 'pancake'

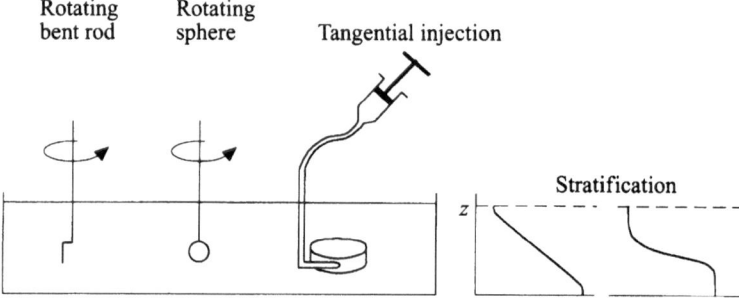

Fig. 1.14 Forcing devices for generation of vortices in a stratified fluid (from [10])

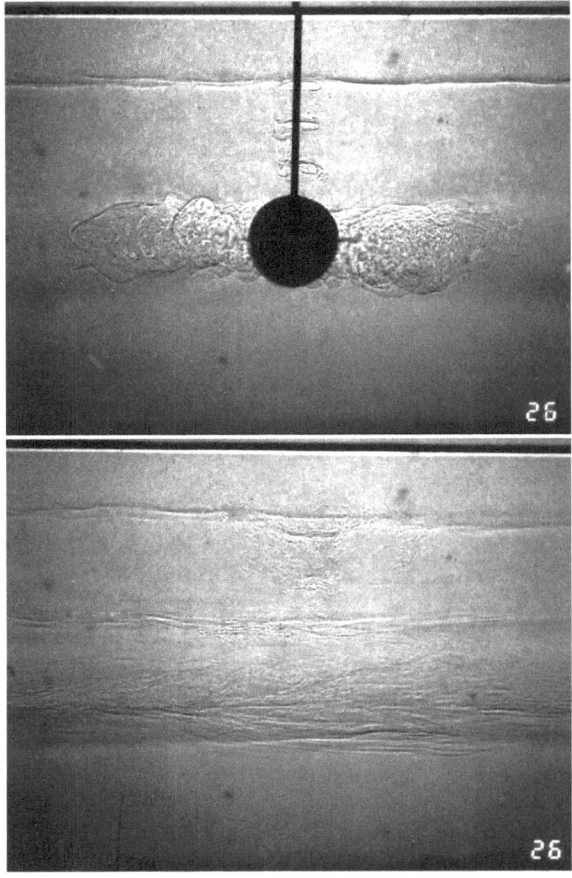

Fig. 1.15 Shadowgraph visualization of the flow generated by a rotating sphere (**a**) during the forcing and (**b**) at $t \simeq 3 \, s$ after the removal of the sphere. Experimental parameters: forcing rotation speed 675 rpm, forcing time 60 s, $N = 1.11$ rad/s, and sphere diameter 3.8 cm (from [11])

shape, with the vertical size of the swirling fluid region being much smaller than its horizontal size L (Fig. 1.16). This implies large gradients of the flow in the z-direction and hence the presence of a radial vorticity component ω_r. Although the swirling motion in these thin vortices is in good approximation planar, the significant vertical gradients imply that the vortex motion is not 2D. Additionally, the strong gradients in z-direction imply a significant effect of diffusion of vorticity in that direction.

Alternatively, a vortex may be generated by tangential injection of fluid in a thin-walled, bottomless cylinder, as also shown in Fig. 1.14. The swirling fluid volume is released by lifting the cylinder vertically. After some adjustment, again a pancake-like vortex is observed with features quite similar to the vortices produced with the spinning devices.

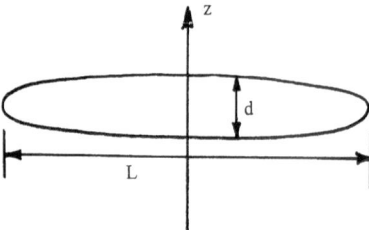

Fig. 1.16 Sketch of the pancake-like structure of the swirling region in the stratified fluid

1.2.3 Decay of Vortices

Flór and van Heijst [11] have measured the velocity distributions in the horizontal symmetry plane for vortices generated by either of the forcing techniques mentioned above. An example of the measured radial distributions of the azimuthal velocity $v_\theta(r)$ and the vertical component ω_z of the vorticity is shown in Fig. 1.17. Since the profiles are scaled by their maximum values V_{max} and ω_{max}, it becomes apparent that the profiles are quite similar during the decay process. This remarkable behaviour motivated Flór and van Heijst [11] to develop a diffusion model that describes viscous diffusion of vorticity in the z-direction. This model was later extended by Trieling and van Heijst [24], who considered diffusion of ω_z from the midplane $z = 0$ (horizontal symmetry plane) in vertical as well as in radial direction. The basic assumptions of this extended diffusion model are the following:

- the midplane $z = 0$ is a symmetry plane;
- at the midplane $z = 0$: $\boldsymbol{\omega} = (0, 0, \omega_z)$;

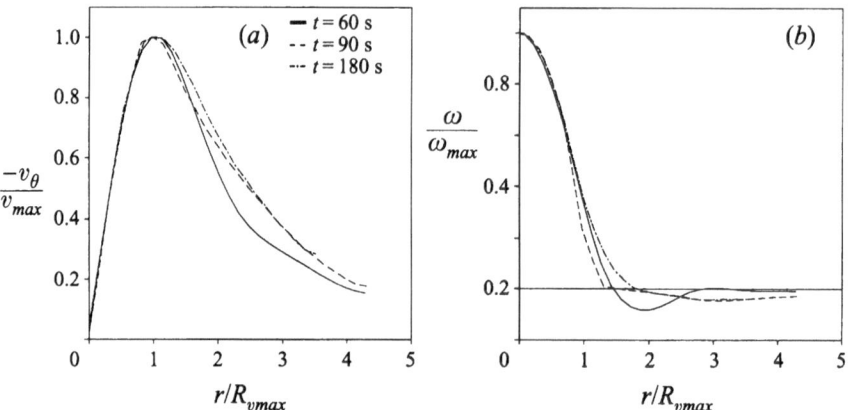

Fig. 1.17 Radial distributions of (**a**) the azimuthal velocity $v_\theta(r)$ and (**b**) the vertical vorticity component ω measured at half-depth in a sphere-generated vortex for three different times t. The profiles have been scaled by the maximum velocity V_{max} and the maximum vorticity ω_x and the radius by the radial position R_{max} of the maximum velocity (from [11])

- near the midplane the evolution of the vertical vorticity ω_z is governed by

$$\frac{\partial \omega_z}{\partial t} + J(\omega_z, \psi) = \nu \nabla_h^2 \omega_z + \nu \frac{\partial^2 \omega_z}{\partial z^2}; \tag{1.57}$$

- axisymmetry implies $J(\omega_z, \psi) = 0$;
- the solution can be written as

$$\omega_z(r, z, t) = \omega(r, t)\Phi(z, t) . \tag{1.58}$$

After substitution of (1.58) in (1.57) one arrives at

$$\frac{\partial \omega}{\partial t} = \frac{\nu}{r}\frac{\partial}{\partial r}\left(r\frac{\partial \omega}{\partial r}\right) , \tag{1.59}$$

$$\frac{\partial \Phi}{\partial t} = \nu \frac{\partial^2 \Phi}{\partial z^2} . \tag{1.60}$$

Apparently, the horizontal diffusion and the vertical diffusion are separated, as they are described by two separate equations. For an isolated vortex originally concentrated in one singular point, Taylor [23] derived the following solution for the *horizontal diffusion* equation (1.59):

$$\omega(r, t) = \frac{C}{(\nu t)^2}\left[1 - \frac{r^2}{4\nu t}\right]\exp\left(-\frac{r^2}{4\nu t}\right) . \tag{1.61}$$

Since we are considering radial diffusion of a non-singular initial vorticity distribution, this solution is modified and written as

$$\omega(r, t) = \frac{C}{\nu^2(t + t_0)^2}\left[1 - \frac{r^2}{4\nu(t + t_0)}\right]\exp\left(-\frac{r^2}{4\nu(t + t_0)}\right) . \tag{1.62}$$

The corresponding expression of the azimuthal velocity is

$$v_\theta(r, t) = \frac{Cr}{2\nu^2(t + t_0)^2}\exp\left(-\frac{r^2}{4\nu(t + t_0)}\right) . \tag{1.63}$$

From this solution it appears that the radius r_m of the peak velocity v_{max} is given by

$$r_m^2 = r_0^2 + 2\nu t , \quad \text{with } r_0 = \sqrt{2\nu t_0} . \tag{1.64}$$

After introducing the following scaling:

$$\tilde{r} = r/r_m , \quad \tilde{\omega} = \omega/\omega_m , \quad \tilde{v}_\theta = v_\theta/\omega_m r_m, \tag{1.65}$$

the solutions (1.62) and (1.63) can be written as

$$\tilde{\omega} = \left[1 - \tfrac{1}{2}\tilde{r}^2\right] \exp\left(-\tfrac{1}{2}\tilde{r}^2\right) \tag{1.66}$$

$$\tilde{v}_\theta = \tfrac{1}{2}\tilde{r} \exp\left(-\tfrac{1}{2}\tilde{r}^2\right) . \tag{1.67}$$

This scaled solution reveals a 'Gaussian vortex', although changing in time.
In order to solve (1.60) for the *vertical diffusion*, the following initial condition is assumed:

$$\Phi(z, 0) = \Phi_0 \cdot \delta(z) , \tag{1.68}$$

with $\delta(z)$ the Dirac function. The solution of this problem is standard, yielding

$$\Phi(z, t) = \frac{\Phi_0}{\sqrt{\nu t}} \exp\left(-\frac{z^2}{4\nu t}\right) . \tag{1.69}$$

The total solution of the extended diffusion model is then given by

$$\hat{\omega}(r, z, t) = \omega(r, t)\frac{\Phi_0}{\sqrt{\nu t}} \exp\left(-\frac{z^2}{4\nu t}\right) , \tag{1.70}$$

with $\omega(r, t)$ given by (1.62).
According to this result, the decay of the maximum value $\hat{\omega}_{\max}$ of the vertical vorticity component (at $r = 0$) at the halfplane $z = 0$ behaves like

$$\hat{\omega}_{\max} = \frac{C\Phi_0}{\nu^{5/2}(t + t_0)^2 \sqrt{t}} . \tag{1.71}$$

An experimental verification of these results was undertaken by Trieling and van Heijst [25]. Accurate flow measurements in the midplane $z = 0$ of vortices produced by either the spinning sphere or the tangential-injection method showed a very good agreement with the extended diffusion model, as illustrated in Fig. 1.18. The agreement of the data points at three different stages of the decay process corresponds excellently with the Gaussian-vortex model (1.66) and (1.67). Also the time evolutions of other quantities like r_m, ω_m, and v_m/r_m show a very good correspondence with the extended diffusion model. For further details, the reader is referred to [25]. In order to investigate the vertical structure of the vortices produced by the tangential-injection method, Beckers et al. [2] performed flow measurements at different horizontal levels. These measurements confirmed the z-dependence according to (1.70). Their experiments also revealed a remarkable feature of the vertical distribution of the density ρ, see Fig. 1.19.

Just after the tangential injection, the density profile shows more or less a two-layer stratification within the confining cylinder, with a relatively sharp interface between the upper and the lower layers. During the subsequent evolution of the vortex after removing the cylinder, this sharp gradient vanishes gradually. In order to better understand the effect of the density distribution on the vortex dynamics, we

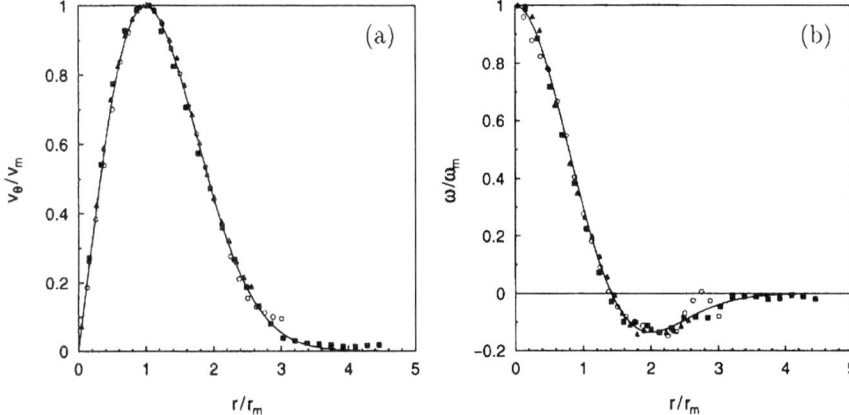

Fig. 1.18 Scaled profiles of (**a**) the azimuthal velocity and (**b**) the vertical vorticity of a vortex generated by the spinning sphere. The measured profiles correspond to three different times: $t = 120$ s (*squares*), 480 s (*circles*), and 720 s (*triangles*). The *lines* represent the Gaussian-vortex model (1.66)–(1.67) (from [25])

consider the equation of motion. Under the assumption of a dominating azimuthal motion, the non-dimensional r, θ, z-components of the Navier–Stokes equation for an axisymmetric vortex are

$$-\frac{v_\theta^2}{r} = -\frac{\partial p}{\partial r} \tag{1.72}$$

$$\frac{\partial v_\theta}{\partial t} = \frac{1}{Re}\left(\frac{\partial^2 v_\theta}{\partial r^2} + \frac{1}{r}\frac{\partial v_\theta}{\partial r} - \frac{v_\theta}{r^2} + \frac{\partial^2 v_\theta}{\partial z^2}\right) \tag{1.73}$$

$$0 = -\frac{\partial p}{\partial z} - \frac{\rho}{F^2} \tag{1.74}$$

with

$$Re = VL/\nu \qquad \text{Reynolds number}$$
$$F = V/(LN) \qquad \text{Froude number}$$

both based on typical velocity and length scales V and L, respectively. The radial component (1.72) describes the *cyclostrophic* balance – see (1.24). The azimuthal component (1.73) describes diffusion of v_θ in r, z-directions, while the z-component (1.74) represents the hydrostatic balance. Elimination of the pressure in (1.72) and (1.74) yields

$$F^2\frac{2v_\theta}{r}\frac{\partial v_\theta}{\partial z} + \frac{\partial \rho}{\partial r} = 0\,. \tag{1.75}$$

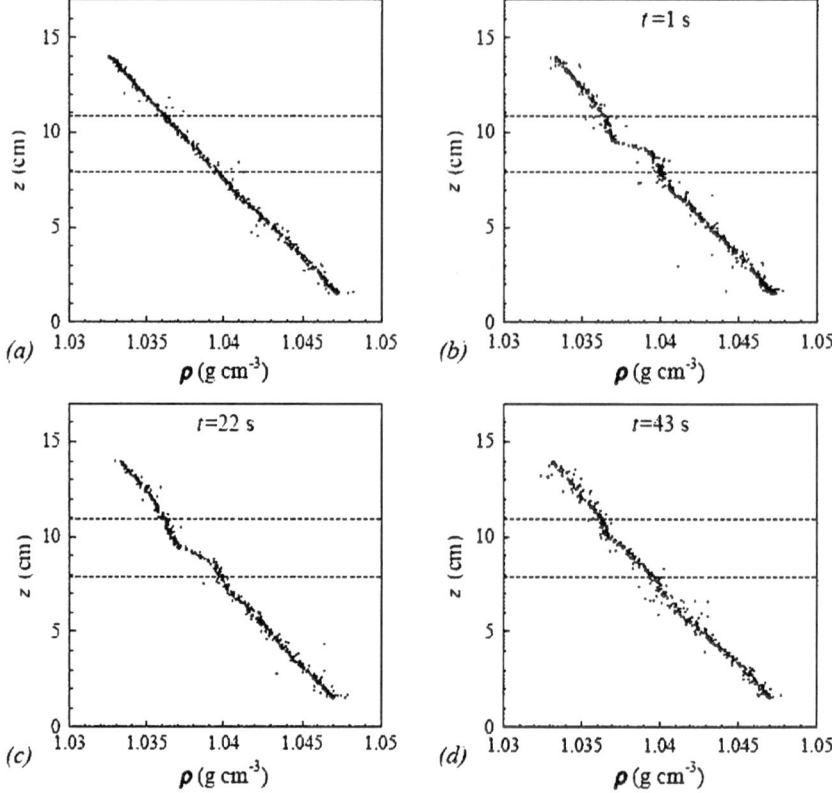

Fig. 1.19 Vertical density structures in the centre of the vortex produced with the tangential-injection method. The profiles are shown (**a**) before the injection, (**b**) just after the injection, but with the cylinder still present, (**c**) soon after the removal of the cylinder, and (**d**) at a later stage (from [2])

This is essentially the 'thermal wind' balance, which relates horizontal density gradients ($\frac{\partial \rho}{\partial r}$) with vertical shear in the cyclostrophic velocity field ($\frac{\partial v_\theta}{\partial z}$). Obviously, the vortex flow field v_θ implies a specific density field to have a cyclostrophically balanced state. In order to study the role of the cyclostrophic balance, numerical simulations based on the full Navier–Stokes equations for axisymmetric flow have been carried out by Beckers et al. [2] for a number of different initial conditions. In case 1 the initial state corresponds with a density perturbation but with $v_\theta = 0$, i.e. without the swirling flow required for the cyclostrophic balance (1.72). The initial state of case 2 corresponds with a swirling flow v_θ, but without the density structure to keep it in the cyclostrophic balance as expressed by (1.75). In both cases, a circulation is set up in the r, z-plane, because either the radial density gradient force is not balanced (case 1) or the centrifugal force is not balanced (case 2). Figure 1.20 shows schematic drawings of the resulting circulation in the r, z-plane for both cases. A circulation in the r, z-plane implies velocity components v_r and v_z, and hence an azimuthal vorticity component ω_θ, defined as

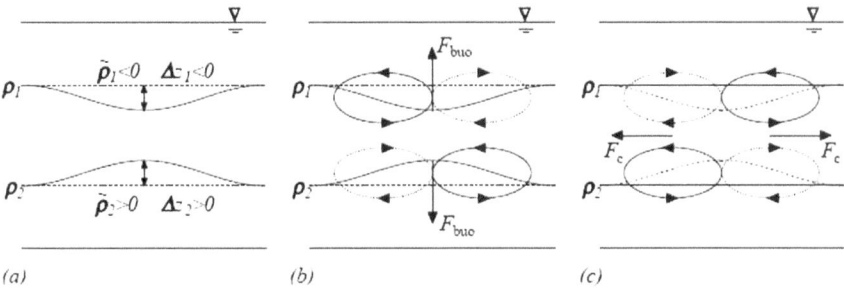

(a)   *(b)*   *(c)*

Fig. 1.20 (a) Schematic drawing of the shape of two isopycnals corresponding with the density perturbation introduced in case 1, with the resultant circulation sketched in (**b**). The resulting circulation arising in case 2, in which the centrifugal force is initially not in balance with the radial density gradient, is shown in (**c**) (from [2])

$$\omega_\theta = \frac{\partial v_r}{\partial z} - \frac{\partial v_z}{\partial r} . \tag{1.76}$$

The numerically calculated spatial and temporal evolutions of ω_θ as well as the density perturbation $\tilde\rho$ are shown graphically in Fig. 1.21. Soon after the density perturbation is released, a double cell circulation pattern is visible in the ω_θ plot, accompanied by two weaker cells. The multiple cells in the later contour plots indicate the occurrence of internal waves radiating away from the origin. A similar behaviour can be observed for case 2, see Fig. 1.22. Additional simulations were carried out for an initially balanced vortex (case 3). In this case the simulations do not show any pronounced waves – as is to be expected for a balanced vortex. How-

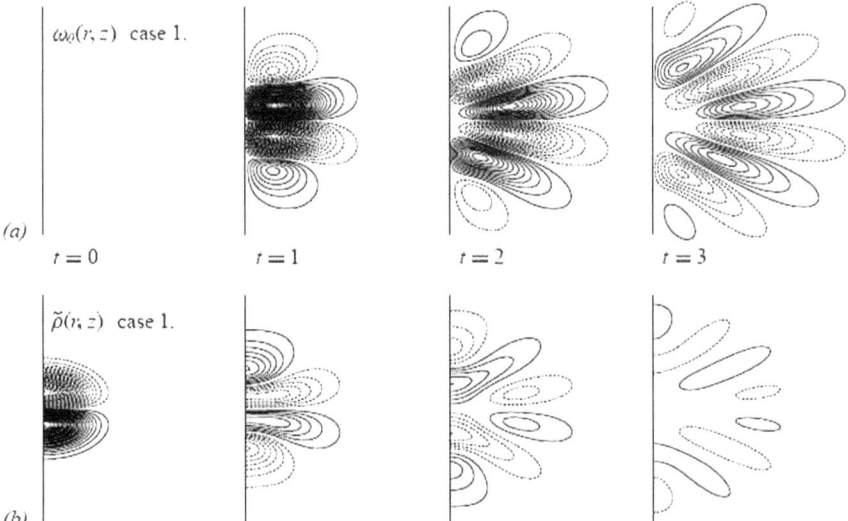

Fig. 1.21 Contour plots in the r, z-plane of the azimuthal vorticity ω_θ in (**a**) and the density perturbation $\tilde\rho$ in (**b**) as simulated numerically for case 1 (from [2])

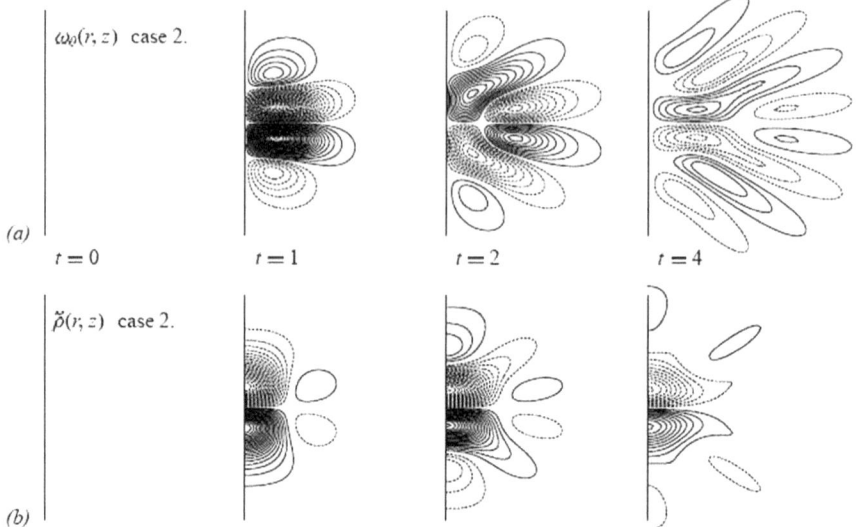

Fig. 1.22 Similar as Fig. 1.21, but now for numerical simulation case 2 (from [2])

ever, due to diffusion the velocity structure changes slowly in time, thus bringing the vortex slightly out of balance. As a result, the flow system adjusts, giving rise to the formation of weak ω_θ and changes in $\tilde{\rho}$. For further details, the reader is referred to Beckers et al. [2].

1.2.4 Instability and Interactions

The pancake-shaped vortices described above may under certain conditions become unstable. For example, the monopolar vortex can show a transition into a tripolar structure, as described by Flór and van Heijst [11]. A more detailed experimental and numerical study was performed by Beckers et al. [3], which revealed that the tripole formation critically depends on the values of the Reynolds number Re, the Froude number F, and the 'steepness' of the initial azimuthal velocity profile. In addition to the tripolar vortex, which can be considered as a wavenumber $m = 2$ instability of the monopolar vortex, higher-order instability modes were formed in specially designed experiments by Beckers [1]. In these experiments the vortices were generated by the tangential-injection method, but now in the annular region between two thin-walled cylinders, thus effectively increasing the steepness of the v_θ profile of the released vortex and promoting higher-order instability modes. Besides, $m = 3$ and $m = 4$ instability was promoted by adding perturbations of this wavenumber in the form of thin metal strips connecting inner and outer cylinders under angles of $120°$ and $90°$, respectively. In the former case the monopolar vortex

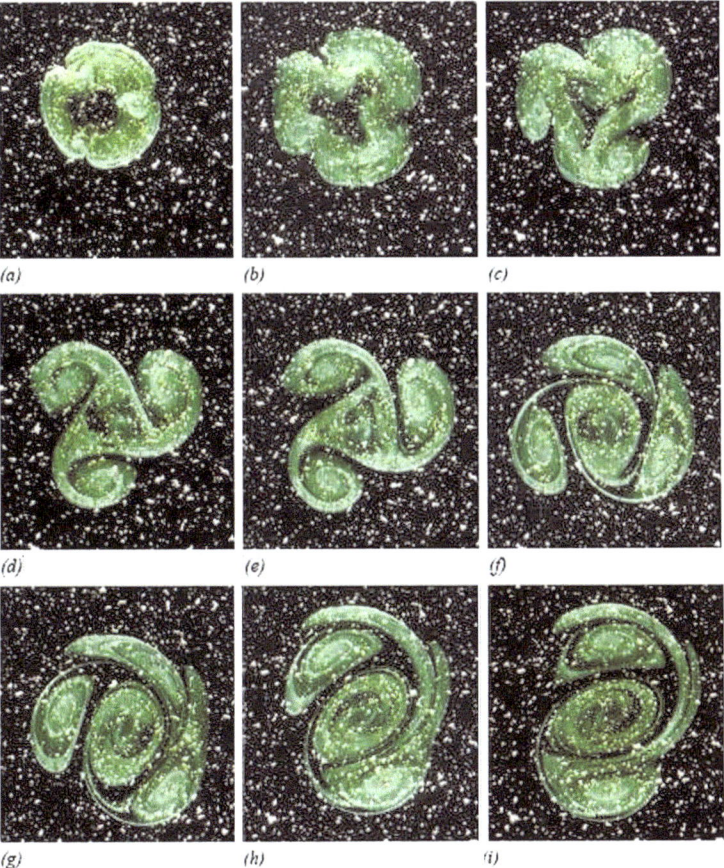

Fig. 1.23 Sequence of dye-visualization pictures showing the evolution of a pancake-shaped vortex in a stratified fluid on which an $m = 4$ perturbation was imposed (from [1])

quickly transformed into a triangular core vortex with three counter-rotating satellite vortices at its sides. This structure turned out to be unstable and was observed to show a transition to a stable tripolar structure. Likewise, the $m = 4$ perturbation led to the formation of a square core vortex with four satellite vortices at its sides. Again, this vortex structure was unstable, showing a quick transformation into a tripolar vortex, with the satellite vortices at relative large separation distances from the core vortex, see Fig. 1.23. Details of these experiments can be found in Appendix A of the PhD thesis of Beckers [1]. It should be noted that this type of instability behaviour was also found in the numerical/experimental study of Kloosterziel and Carnevale [19] on 2D vortices in a rotating fluid.

Schmidt et al. [26] investigated the interaction of monopolar, pancake-like vortices generated close to each other, on the same horizontal level. In their experiments,

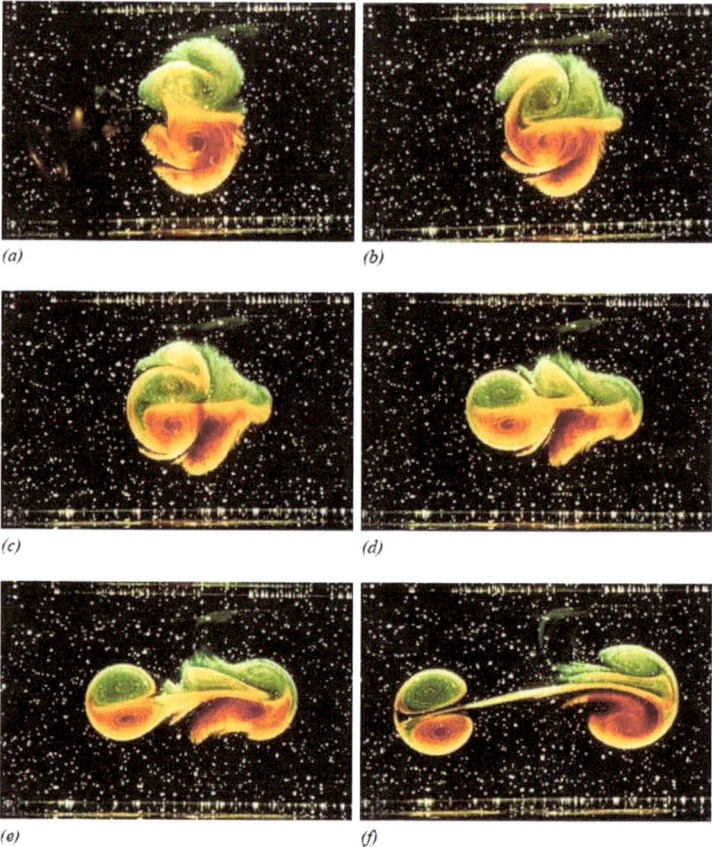

(a)   (b)

(c)   (d)

(e)   (f)

Fig. 1.24 Sequence of dye-visualization pictures showing the evolution of two counter-rotating pancake vortices released at small separation distance (from [26])

they created vortices by the tangential-injection method, while they systematically changed the distance between the confining cylinders. A remarkable result was obtained for counter-rotating vortices at the closest possible separation distance, viz. with the cylinders touching. After vertically withdrawing the cylinders the vortices showed an interesting behaviour, shown by the sequence of dye-visualization pictures displayed in Fig. 1.24. Apparently, the two monopolar vortices finally give rise to two dipole structures moving away from each other. The explanation for this behaviour lies in the fact that the vortices generated with this tangential-injection device are 'isolated', i.e. their net circulation is zero (because of the no-slip condition at the inner cylinder wall): each vortex has a vorticity core surrounded by a ring of oppositely signed vorticity. The dye visualization clearly shows that the cores quickly combine into one dipolar vortex, while the shields of opposite vorticity are advected forming a second, weaker dipolar vortex moving in opposite direction.

1.3 Concluding Remarks

In the preceding sections we have discussed some basic dynamical features of vortices in rotating fluids (Sect. 1.1) and stratified fluids (Sect. 1.2). By way of illustration of the theoretical issues, a number of laboratory experiments on vortices were highlighted. Given the scope of this chapter, we had to restrict ourselves in the discussion and the selection and presentation of the material was surely biased by the author's involvement in a number of studies of this type of vortices. For example, much more can be said about vortex instability. What about the dynamics of tall vortices in a stratified fluid? What about interactions of pancake-shaped vortices generated at different levels in the stratified fluid column? Some of these questions will be treated in more detail by Chomaz et al. [8] in Chapter 2 of this volume.

Other interesting phenomena can be encountered when rotation and stratification are present simultaneously. In that case, the structure and shape of coherent vortices are highly dependent on the ratio f/N, see, e.g. Reinaud et al. [22]. These and many more aspects of geophysical vortex dynamics fall outside the limited scope of this introductory text.

Acknowledgments The author gratefully acknowledges Jan-Bert Flór and his colleagues for having organized the summer school on 'Fronts, Waves, and Vortices' in 2006 in the Valsavarenche mountain valley near Aosta, Italy.

References

1. Beckers, M.: Dynamics of vortices in a stratified fluid. Ph.D. thesis, Eindhoven University of Technology (1999). http://www.fluid.tue.nl/pub/index.html.
2. Beckers, M., Verzicco, R., Clercx, H.J.H., van Heijst, G.J.F.: Dynamics of pancake-like vortices in a stratified fluid: experiments, model and numerical simulations. J. Fluid Mech. **433**, 1–27 (2001).
3. Beckers, M., Clercx, H.J.H., van Heijst, G.J.F., Verzicco, R.: Evolution and instability of monopolar vortices in a stratified fluid. Phys. Fluids **15**, 1033–1045 (2003).
4. Cariteau, B., Flór, J.B.: Experimental study on columnar vortex interactions in rotating stratified fluids, J. Fluid Mech. submitted (2010).
5. Carnevale, C.F., Kloosterziel, R.C., van Heijst, G.J.F.: Propagation of barotropic vortices over topography in a rotating tank. J. Fluid Mech. **233**, 119–139 (1991).
6. Carton, J., McWilliams, J.C.: Barotropic and baroclinic instabilities of axisymmetric vortices in a quasi-geostrophic model. In: Nihoul, J.C.J., Jamart, B.M. (eds.) Mesoscopic/Synoptic Coherent Structures in Geophysical Turbulence, pp. 225–244, Elsevier, Amsterdam (1989).
7. Cenedese, C.: Laboratory experiments on mesoscale vortices colliding with a seamount. J. Geophys. Res. **C107**(C6), 3053 (2002).
8. Chomaz, J.M., Ortiz, S., Gallaire, F., Billant, P.: Stability of quasi-two-dimensional vortices. Lect. Notes Phys. **805**, 35–59, Springer, Heidelberg (2010).
9. Flierl, G.R.: On the instability of geostrophic vortices. J. Fluid Mech. **197**, 349–388 (1988).
10. Flór, J.B.: Coherent vortex structures in stratified fluids. Ph.D. thesis, Eindhoven University of Technology (1994).
11. Flór, J.B., van Heijst, G.J.F.: Stable and unstable monopolar vortices in a stratified fluid. J. Fluid Mech. **311**, 257–287 (1996).

12. Griffiths, R.W., Linden, P.F.: The stability of vortices in a rotating stratified fluid. J. Fluid Mech. **105**, 283–306 (1981).
13. van Heijst, G.J.F.: Topography effects on vortices in a rotating fluid. Meccanica **29**, 431–451 (1994).
14. van Heijst, G.J.F., Clercx, H.J.H.: Laboratory modeling of geophysical vortices. Annu. Rev. Fluid Mech. **41**, 143–164 (2009).
15. Holton, J.R.: An Introduction to Dynamics Meteorology, 3rd edn. Academic Press, San Diego (1992).
16. Hopfinger, E.J., van Heijst, G.J.F.: Vortices in rotating fluids. Annu. Rev. Fluid Mech. **25**, 241–289 (1993).
17. Kloosterziel, R.C., van Heijst, G.J.F.: An experimental study of unstable barotropic vortices in a rotating fluid. J. Fluid Mech. **223**, 1–24 (1991).
18. Kloosterziel, R.C., van Heijst, G.J.F.: The evolution of stable barotropic vortices in a rotating free-surface fluid. J. Fluid Mech. **239**, 607–629 (1992).
19. Kloosterziel, R.C., Carnevale, G.F.: On the evolution and saturation of instabilities of twodimensional isolated circular vortices. J. Fluid Mech. **388**, 217–257 (1999).
20. Maas, L.R.M.: Nonlinear and free-surface effects on the spin-down of barotropic axisymmetric vortices. J. Fluid Mech. **246**, 117–141 (1993).
21. Meleshko, V.V., van Heijst, G.J.F.: On Chaplygin's investigations of two-dimensional vortex structures in an inviscid fluid. J. Fluid Mech. **272**, 157–182 (1994).
22. Reinaud, J., Dritschel, D.G., Koudella, C.R.: The shape of vortices in quasi-geostropic turbulence. J. Fluid Mech. **474**, 175–192 (2003).
23. Taylor, G.I.: On the dissipation of eddies. In: Batchelor, G.K. (ed.) The Scientific Papers of Sir Geoffrey Ingram Taylor, vol. 2: Meteorology, Oceanography and Turbulent Flow, pp. 96–101. Cambridge University Press, Cambridge (1918).
24. Trieling, R.R., Linssen, A.H., van Heijst, G.J.F.: Monopolar vortices in an irrotational annular shear flow. J. Fluid Mech. **360**, 273–294 (1998).
25. Trieling, R.R., van Heijst, G.J.F.: Decay of monopolar vortices in a stratified fluid. Fluid Dyn. Res. **23**, 27–43 (1998).
26. Schmidt, M.R., Beckers, M., Nielsen, A.H., Juul Rasmussen, J., van Heijst, G.J.F.: On the interaction between oppositely-signed, shielded, monopolar vortices. Phys. Fluids **10**, 3099–3110 (1998).
27. Velasco Fuentes, O.U., van Heijst, G.J.F.: Experimental study of dipolar vortices on a topographic β-plane. J. Fluid Mech. **259**, 79–106 (1994).
28. Velasco Fuentes, O.U., van Heijst, G.J.F., Cremers, B.E.: Chaotic transport by dipolar vortices on a β-plane. J. Fluid Mech. **291**, 139–161 (1995).
29. Whitehead, J.A., Stern, M.E., Flierl, G.R., Klinger, B.A.: Experimental observations of baroclinic eddies on a sloping bottom. J. Geophys. Res. **95**, 9585–9610 (1990).
30. Zabusky, N.J., McWilliams, J.C.: A modulated point-vortex model for geostrophic, β-plane dynamics. Phys. Fluids **25**, 2175–2182 (1982).
31. Zavala Sansón, L., van Heijst, G.J.F., Backx, N.A.: Ekman decay of a dipolar vortex in a rotating fluid. Phys. Fluids **13**, 440–451 (2001).
32. Zavala Sansón, L., van Heijst, G.J.F.: Ekman effects in a rotating flow over bottom topography. J. Fluid Mech. **471**, 239–255 (2002).

Chapter 2
Stability of Quasi Two-Dimensional Vortices

J.-M. Chomaz, S. Ortiz, F. Gallaire, and P. Billant

Large-scale coherent vortices are ubiquitous features of geophysical flows. They have been observed as well at the surface of the ocean as a result of meandering of surface currents but also in the deep ocean where, for example, water flowing out of the Mediterranean sea sinks to about 1000 m deep into the Atlantic ocean and forms long-lived vortices named Meddies (Mediterranean eddies). As described by Armi et al. [1], these vortices are shallow (or pancake): they stretch out over several kilometers and are about 100 m deep. Vortices are also commonly observed in the Earth or in other planetary atmospheres. The Jovian red spot has fascinated astronomers since the 17th century and recent pictures from space exploration show that mostly anticyclonic long-lived vortices seem to be the rule rather than the exception. For the pleasure of our eyes, the association of motions induced by the vortices and a yet quite mysterious chemistry exhibits colorful paintings never matched by the smartest laboratory flow visualization (see Fig. 2.1). Besides this decorative role, these vortices are believed to structure the surrounding turbulent flow. In all these cases, the vortices are large scale in the horizontal direction and shallow in the vertical. The underlying dynamics is generally believed to be two-dimensional (2D) in first approximation. Indeed both the planetary rotation and the vertical strong stratification constrain the motion to be horizontal. The motion tends to be uniform in the vertical in the presence of rotation effects but not in the presence of stratification. In some cases the shallowness of the fluid layer also favors the two-dimensionalization of the vortex motion. In the present contribution, we address the following question: Are such coherent structures really 2D? In order to do so, we discuss the stability of such structures to three-dimensional (3D) perturbations paying particular attention to the timescale and the length scale on which they develop. Five instability mechanisms will be discussed, all having received renewed attention in the past few years. The shear instability and the generalized centrifugal instability apply to isolated vortices. Elliptic and hyperbolic instability involve an extra straining effect due to surrounding vortices or to mean shear. The newly discovered zigzag instability

J.-M. Chomaz, S. Ortiz, F. Gallaire, P. Billant
Ladhyx, CNRS-École polytechnique, 91128 Palaiseau, France,
chomaz@ladhyx.polytechnique.fr

Chomaz, J.-M. et al.: *Stability of Quasi Two-Dimensional Vortices*. Lect. Notes Phys. **805**, 35–59 (2010)
DOI 10.1007/978-3-642-11587-5_2
© Springer-Verlag Berlin Heidelberg 2010

Fig. 2.1 Artwork by Ando Hiroshige

also originates from the straining effect due to surrounding vortices or to mean shear, but is a "displacement mode" involving large horizontal scales yet small vertical scales.

2.1 Instabilities of an Isolated Vortex

Let us consider a vertical columnar vortex in a fluid rotating at angular velocity Ω in the presence of a stable stratification with a Brunt–Väisälä frequency $N^2 = \frac{d \ln \rho}{dz} g$. The vortex is characterized by a distribution of vertical vorticity, ζ_{max}, which, from now on, only depends on the radial coordinate r and has a maximum value η_{max}. The flow is then defined by two nondimensional parameters: the Rossby number $Ro = \frac{\zeta_{max}}{2\Omega}$ and the Froude number $F = \frac{\zeta_{max}}{N}$. The vertical columnar vortex is first assumed to be axisymmetric and isolated from external constrains. Still it may exhibit two types of instability, the shear instability and the generalized centrifugal instability.

2.1.1 The Shear Instability

The vertical vorticity distribution exhibits an extremum:

$$\frac{d\zeta}{dr} = 0. \tag{2.1}$$

Rayleigh [44] has shown that the configuration is potentially unstable to the Kelvin–Helmholtz instability. This criterion is similar to the inflexional velocity profile criterion for planar shear flows (Rayleigh [43]). These modes are 2D and therefore insensitive to the background rotation. They affect both cyclones and anticyclones and only depend on the existence of a vorticity maximum or minimum at a certain radius. As demonstrated by Carton and McWilliams [11] and Orlandi and Carnevale [36] the smaller the shear layer thickness, the larger the azimuthal wavenumber m that is the most unstable. Three-dimensional modes with low axial wavenumber are also destabilized by shear but their growth rate is smaller than in the 2D limit. This instability mechanism has been illustrated by Rabaud et al. [42] and Chomaz et al. [13] (Fig. 2.2).

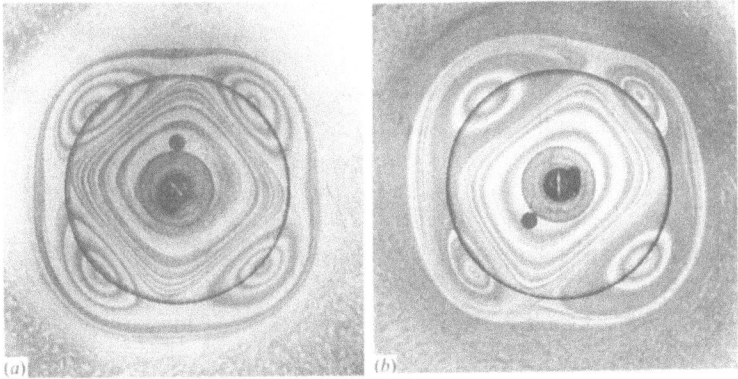

Fig. 2.2 Azimuthal Kelvin–Helmholtz instability as observed by Chomaz et al. [13]

2.1.2 The Centrifugal Instability

In another famous paper, Rayleigh [45] also derived a sufficient condition for stability, which was extended by Synge [47] to a necessary condition in the case of axisymmetric disturbances. This instability mechanism is due to the disruption of the balance between the centrifugal force and the radial pressure gradient. Assuming that a ring of fluid of radius r_1 and velocity $u_{\theta,1}$ is displaced at radius r_2 where the velocity equals $u_{\theta,2}$, (see Fig. 2.3) the angular momentum conservation implies that it will acquire a velocity $u'_{\theta,1}$ such that $r_1 u_{\theta,1} = r_2 u'_{\theta,1}$. Since the ambient pressure gradient at r_2 exactly balances the centrifugal force associated to a velocity $u_{\theta,2}$, it amounts to $\partial p/\partial r = \rho u_{\theta,2}^2/r_2$. The resulting force density at $r = r_2$ is $\frac{\rho}{r_2}((u_{\theta,1}')^2 - (u_{\theta,2})^2)$. Therefore, if $(u_{\theta,1}')^2 < (u_{\theta,2})^2$, the pressure gradient overcomes the angular momentum of the ring which is forced back to its original position, while if on contrary $(u_{\theta,1}')^2 > (u_{\theta,2})^2$, the situation is unstable. Stability is therefore ensured if $u_{\theta,1}^2 r_1^2 < u_{\theta,2}^2 r_2^2$. The infinitesimal analog of this reasoning yields the Rayleigh instability criterion

$$\frac{d}{dr}(u_\theta r)^2 \le 0, \tag{2.2}$$

or equivalently

$$\delta = 2\zeta u_\theta/r < 0, \tag{2.3}$$

where ζ indicates the axial vorticity and δ is the so-called Rayleigh discriminant. In reality, the fundamental role of the Rayleigh discriminant was further understood through Bayly's [2] detailed interpretation of the centrifugal instability in the context of so-called shortwave stability theory, initially devoted to elliptic and hyperbolic instabilities (see Sect. 2.3 and Appendix). Bayly [2] considered non-axisymmetric flows, with closed streamlines and outward diminishing circulation. He showed that the negativeness of the Rayleigh discriminant on a whole closed streamline implied the existence of a continuum of strongly localized unstable eigenmodes for which pressure contribution plays no role. In addition, it was shown that the most unstable mode was centered on the radius r_{min} where the Rayleigh discriminant reaches its negative minimum $\delta(r_{min}) = \delta_{min}$ and displayed a growth rate equal to $\sigma = \sqrt{-\delta_{min}}$.

On the other hand, Kloosterziel and van Heijst [21] generalized the classical Rayleigh criterion (2.3) in a frame rotating at rate Ω for circular streamlines. This centrifugal instability occurs when the fluid angular momentum decreases outward:

$$2r^3 \frac{d\left(r^2(\Omega + u_\theta/r)\right)^2}{dr} = (\Omega + u_\theta/r)\,(2\Omega + \zeta) < 0. \tag{2.4}$$

This happens as soon as the absolute vorticity $\zeta + 2\Omega$ or the absolute angular velocity $\Omega + u_\theta/r$ changes sign. If vortices with a relative vorticity of a single

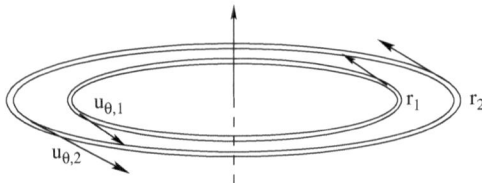

Fig. 2.3 Rayleigh centrifugal instability mechanism

sign are considered, centrifugal instability may occur only for anticyclones when the absolute vorticity is negative at the vortex center, i.e., if Ro^{-1} is between -1 and 0. The instability is then localized at the radius where the generalized Rayleigh discriminant reaches its (negative) minimum.

In a rotating frame, Sipp and Jacquin [48] further extended the generalized Rayleigh criterion (2.4) for general closed streamlines by including rotation in the framework of shortwave stability analysis, extending Bayly's work. A typical example of the distinct cyclone/anticyclone behavior is illustrated in Fig. 2.4 where a counter-rotating vortex pair is created in a rotating tank (Fontane [19]). For this value of the global rotation, the columnar anticyclone on the right is unstable while the cyclone on the left is stable and remains columnar. The deformations of the anticyclone are observed to be axisymmetric rollers with opposite azimuthal vorticity rings.

The influence of stratification on centrifugal instability has been considered to further generalize the Rayleigh criterion (2.4). In the inviscid limit, Billant and Gallaire [9] have shown the absence of influence of stratification on large wavenumbers: a range of vertical wavenumbers extending to infinity are destabilized by the centrifugal instability with a growth rate reaching asymptotically $\sigma = \sqrt{-\delta_{\min}}$. They also showed that the stratification will re-stabilize small vertical wavenumbers but leave unaffected large vertical wavenumbers. Therefore, in the inviscid stratified case, axisymmetric perturbations with short axial wavelength remain the most unstable, but when viscous effects are, however, also taken into account, the leading

Anticyclone   Cyclone

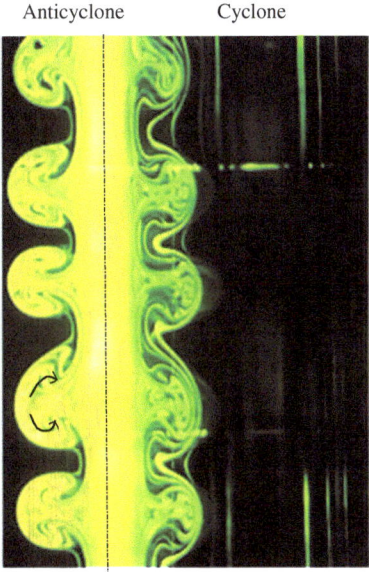

Fig. 2.4 Centrifugal instability in a rotating tank. The columnar vortex on the left is an anticyclone and is centrifugally unstable whereas the columnar vortex on the right is a stable cyclone (Fontane [19])

unstable mode becomes spiral for particular Froude and Reynolds number ranges (Billant et al. [7]).

2.1.3 Competition Between Centrifugal and Shear Instability

Rayleigh's criterion is valid for axisymmetric modes ($m = 0$). Recently Billant and Gallaire [9] have extended the Rayleigh criterion to spiral modes with any azimuthal wave number m and derived a sufficient condition for a free axisymmetric vortex with angular velocity u_θ/r to be unstable to a three-dimensional perturbation of azimuthal wavenumber m: the real part of the growth rate

$$\sigma(r) = -imu_\theta/r + \sqrt{-\delta(r)}$$

is positive at the complex radius $r = r_0$ where $\partial\sigma(r)/\partial r = 0$, where $\delta(r) = (1/r^3)\partial(r^2 u_\theta^2)/\partial r$ is the Rayleigh discriminant. The application of this new criterion to various classes of vortex profiles showed that the growth rate of non-axisymmetric disturbances decreased as m increased until a cutoff was reached. Considering a family of unstable vortices introduced by Carton and McWilliams [11] of velocity profile $u_\theta = r \exp(-r^\alpha)$, Billant and Gallaire [9] showed that the criterion is in excellent agreement with numerical stability analyses. This approach allows one to analyze the competition between the centrifugal instability and the shear instability, as shown in Fig. 2.5, where it is seen that centrifugal instability dominates azimuthal shear instability.

The addition of viscosity is expected to stabilize high vertical wavenumbers, thereby damping the centrifugal instability while keeping almost unaffected

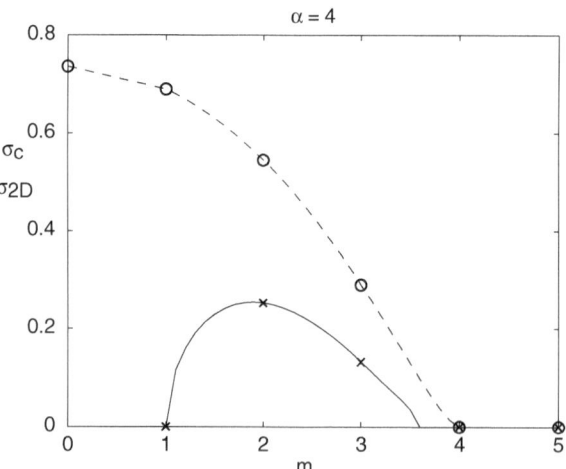

Fig. 2.5 Growth rates of the centrifugal instability for $k = \infty$ (*dashed line*) and shear instability for $k = 0$ (*solid line*) for the Carton and McWilliams' vortices [11] for $\alpha = 4$

two-dimensional azimuthal shear modes of low azimuthal wavenumber. This may result in shear modes to become the most unstable.

2.2 Influence of an Axial Velocity Component

In many geophysical situations, isolated vortices present a strong axial velocity. This is the case for small-scale vortices like tornadoes or dust devils, but also for large-scale vortices for which planetary rotation is important, since the Taylor Proudman theorem imposes that the flow should be independent of the vertical in the bulk of the fluid, but it does not impose the vertical velocity to vanish. In this section, we outline the analysis of [29] and [28] on the modifications brought to centrifugal instability by the presence of an axial component of velocity. As will become clear in the sequel, negative helical modes are favored by this generalized centrifugal instability, when axial velocity is also taken into account.

Consider a vortex with azimuthal velocity component u_θ and axial flow u_z. For any radius r_0, the velocity fields may be expanded at leading order:

$$u_\theta(r) = u_\theta^0 + g_\theta(r - r_0), \tag{2.5}$$

$$u_z(r) = u_z^0 + g_z(r - r_0), \tag{2.6}$$

with $g_\theta = \left.\frac{du_\theta}{dr}\right|_{r_0}$ and $g_z = \left.\frac{du_z}{dr}\right|_{r_0}$. By virtue of Rayleigh's principle (2.2), axisymmetric centrifugal instability will prevail in absence of axial flow when

$$\frac{g_\theta r_0}{u_\theta^0} < -1, \tag{2.7}$$

thereby leading to the formation of counter-rotating vortex rings.

When a nonuniform axial velocity profile is present, Rayleigh's argument based on the exchange of rings at different radii should be extended by considering the exchange of spirals at different radii. In that case, these spirals should obey a specific kinematic condition in order for the axial momentum to remain conserved as discussed in [29]. Following his analysis, let us proceed to a change of frame considering a mobile frame of reference at constant but yet arbitrary velocity $\overline{u}$ in the z direction. The flow in this frame of reference is characterized by a velocity field $\tilde{u}_z^0$ such that

$$\tilde{u}_z^0 = u_z^0 - \overline{u}. \tag{2.8}$$

The choice of $\overline{u}$ is now made in a way that the helical streamlines have a pitch which is independent of r in the vicinity of r_0. The condition on $\overline{u}$ is therefore that the distance traveled at velocity $\tilde{u}_z^0$ during the time $\frac{2\pi r_0}{u_\theta^0}$ required to complete an entire revolution should be independent of a perturbation δr of the radius r:

$$\frac{(\tilde{u}_z^0)(2\pi r_0)}{u_\theta^0} = \frac{(\tilde{u}_z^0 + g_z\delta r)(2\pi(r_0 + \delta r))}{u_\theta^0 + g_\theta\delta r}. \tag{2.9}$$

Retaining only dominant terms in δr, this defines a preferential helical pitch α for streamlines in r_0 in the co-moving reference frame:

$$\tan(\alpha) = \frac{\tilde{u}_z^0}{u_\theta^0} = -\frac{g_z r_0/u_\theta^0}{1 - g_\theta r_0/u_\theta^0}. \tag{2.10}$$

In this case, the stream surfaces defined by these streamlines are helical surfaces of identical geometry defining an helical annular tube. This enables [29] to generalize the Rayleigh mechanism by exchanging two spirals in place of rings conserving mass and angular momentum. The underlying geometrical similarity is ensured by the choice of the axial velocity of the co-moving frame. Neglecting the torsion, the obtained flow is therefore similar to the one studied previously. Indeed, the normal to the osculating plane (so-called binormal) is precessing with respect to the z-axis with constant angle α. Ludwieg [29] then suggests to locally apply the Rayleigh criterium introducing following reduced quantities:

- $r_0^{\text{eff}} = \frac{r_0}{\cos^2\alpha}$, the radius of curvature of the helix,
- $u_\theta^{0,\text{eff}} = \sqrt{(u_\theta^0)^2 + (\tilde{u}_z^0)^2}$, the velocity along streamlines,
- $g_\theta^{\text{eff}} = g_\theta\cos\alpha + g_z\sin\alpha$, the gradient of effective azimuthal velocity.

Figure 2.6 represents an helical surface inscribed on a cylinder of radius r_0 and circular section C'. The osculating circle C and osculating plane P containing the tangent and normal are also shown. The application of the Rayleigh criterion yields

$$\frac{g_\theta^{\text{eff}} r_0^{\text{eff}}}{u_\theta^{0,\text{eff}}} = \frac{r_0(g_\theta + g_z\tan\alpha)}{u_\theta^0} < -1. \tag{2.11}$$

Using the value of $\tan\alpha$ (2.10), one is left with

$$\frac{g_\theta r_0}{u_\theta^0} - \frac{(g_z r_0/u_\theta^0)^2}{1 - g_\theta r_0/u_\theta^0} < -1, \tag{2.12}$$

which was found a quarter century after by Leibovich and Stewartson [28], using a completely different and more rigorous method.

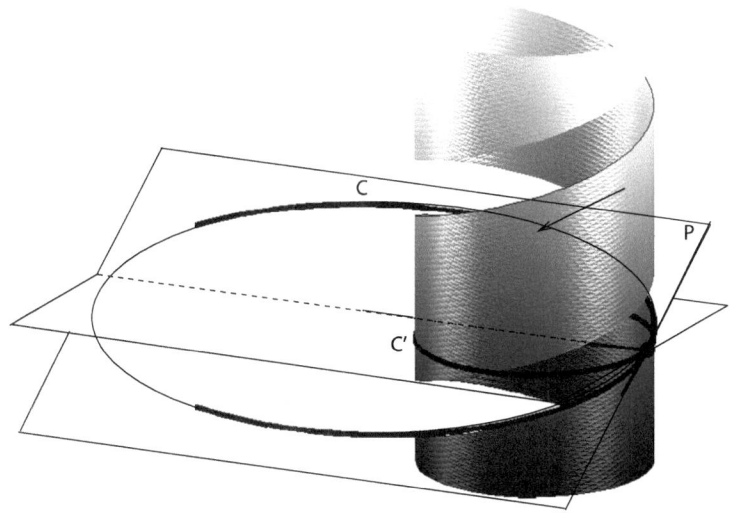

Fig. 2.6 Spiral centrifugal instability mechanism according to [29]

Ludwieg [29] thereby anticipated by physical arguments the asymptotic criterion recovered rigorously by Leibovich and Stewartson [28] showing that, when (2.11) holds, the most unstable helices have a pitch:

$$\tan(\alpha) = k/m = -\frac{g_z r_0 / u_\theta^0}{1 - g_\theta r_0 / u_\theta^0}. \tag{2.13}$$

This result was also derived independently in the shortwave asymptotics WKB framework by Eckhoff and Storesletten [18] and Eckhoff [17]. More recently, following the derivation of Bayly [2], LeBlanc and Le Duc [25] have shown how to construct highly localized modes using the WKB description.

2.3 Instabilities of a Strained Vortex

In a majority of flows, vortices are never isolated but interact one with each other. They may also interact with a background shear imposed by zonal flow like in the Jovian bands. At leading order, this interaction results in a 2D strain field, ϵ, acting on the vortex and more generally on the vorticity field (ϵ and $-\epsilon$ are the eigenvalues of the symmetric part of the velocity gradient tensor, the base flow being assumed 2D). The presence of this strain induces two types of small-scale instability.

2.3.1 The Elliptic Instability

Due to the action of the strain field, the vertical columnar vortex is no more axisymmetric but it takes a steady (or quasi-steady) elliptic shape characterized by elliptic
streamlines in the vortex core (Fig. 2.7). Following the early works of Moore and
Saffman [35], Tsai and Widnall [50], Pierrehumbert [41] Bayly [3], and Waleffe
[50] a tremendous number of studies have shown that the strain field induces a so-
called elliptic instability that acts at all scales. Readers are referred to the reviews by
Cambon [10] and Kerswell [20] for a comprehensive survey of the literature. Here,
we shall develop only the local point of view since it gives insights on the instability
mechanism and on the effect of stratification and rotation.

For a steady basic flow, with elliptical streamlines, Miyazaki [34] analyzed the
influence of a Coriolis force and a stable stratification. The shortwave perturbations
are characterized by a wave vector $\mathbf{k}$ and an amplitude vector $\mathbf{a}$. These lagrangian
Fourier modes, called also Kelvin waves, satisfy the Euler equations under the
Boussinesq approximation (see Appendix for detailed calculation without stratification). Following Lifschitz and Hameiri [33], the flow is unstable if there exists a
streamline on which the amplitude $\mathbf{a}$ is unbounded at large time.

The system evolving along closed trajectories is periodic, and stability may be tack-
led by Floquet analysis. In the particular case of small strain, Leblanc [23], following Waleffe [50], gives a physical interpretation of elliptical instability in terms of
the parametric excitation of inertial waves in the core of the vortex. The instability
problem reduces to a Mathieu equation (2.50) (see Sect. 2.7.5), parametric excitations are found to occur for

$$\left(\frac{\zeta^2}{4}\right) j^2 \; = \; N^2 \sin^2\theta \; + \; (\zeta + 2\Omega)^2 \cos^2\theta, \qquad (2.14)$$

where θ is the angle between the wave vector $\mathbf{k}$ and the spanwise unit vector and
j is an integer. Without stratification and rotation, we retrieve for $j = 1$, that, at

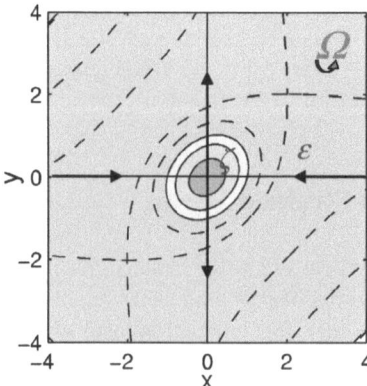

Fig. 2.7 Flow around an elliptic fixed point

small strain, the resonant condition (2.14) is fulfilled only for an angle of $\pi/3$ as demonstrated by Waleffe [50].

For a strain which is not small, the Floquet problem is integrated numerically. Extending Craik's work [15] Miyazaki [34] observed that the classical subharmonic instability of Pierrehumbert [41] and Bayly [3] ($j = 1$) is suppressed when rotation and stratification effects are added. Other resonances are found to occur. According to the condition (2.14), resonance does not exist when either

$$\frac{\zeta}{2} \; < \; \min\left(N, |\zeta + 2\Omega|\right) \quad \text{or} \quad \frac{\zeta}{2} \; > \; \max\left(N, |\zeta + 2\Omega|\right). \tag{2.15}$$

The vortex is then stable with respect to elliptic instability if (Miyazaki [34])

$$F > 2 \text{ and } -2 < RO < -2/3 \quad \text{or} \quad F < 2 \text{ and } (Ro < -2 \text{ or } Ro > -2/3). \tag{2.16}$$

The instability growth rate is (Kerswell [20])

$$\sigma \; = \; \frac{\epsilon\;(3Ro + 2)^2\left(F^2 - 4\right)}{16\left(F^2(Ro + 1)^2 - 4Ro^2\right)}. \tag{2.17}$$

The flow is then unstable with respect to hyperbolic instability in the vicinity of $Ro = -2$:

$$\frac{-2}{(1 - 2\epsilon/\zeta)} \; < \; Ro \; < \; \frac{-2}{(1 + 2\epsilon/\zeta)}. \tag{2.18}$$

We want to emphasize that a rotating stratified flow is characterized by two timescales N^{-1} and Ω^{-1}. If we consider the effect of a strain field on a uniform vorticity field, two timescales are added ϵ^{-1} and ζ^{-1} but no length scale. This explains why all the modes are destabilized in a similar manner, no matter how large the wave vector is.

Indeed in the frame rotating with the vortex core (i.e., at an angular velocity $\zeta/2 + \Omega$) the Coriolis force acts as a restoring force and is associated with the propagation of inertial waves. When the fluid is stratified, the buoyancy is a second restoring force and modifies the properties of inertial waves, these two effects combine in the dispersion relation for propagating inertial-gravity waves. The local approach has been compared with the global approach by Le Dizès [26] in the case of small strain and for a Lamb–Oseen vortex.

In the frame rotating with the vortex core, the strain field rotates at the angular speed $-\zeta/2$ and since the elliptic deformation is a mode $m = 2$, the fluid in the core of the vortex "feels" consecutive contractions and dilatations at a pulsation $2\zeta/2$ (i.e., twice faster than the strain field). These periodic constrains may destabilize inertial gravity waves via a subharmonic parametric instability when their pulsations equal half the forcing frequency. If the deformation field were tripolar instead of dipolar, the resonance frequency would have been $3\zeta/4$ but the physics would have been the same (Le Dizès and Eloy [27], Eloy and Le Dizès [30]).

Elliptical instability in an inertial frame occurs for oblique wave vectors and thus needs pressure contribution. When rotation is included, for anticyclonic rotation the most unstable wave vector becomes a purely spanwise mode with $\theta = 0$. In that case, the contribution of pressure is not necessary and disappears from the evolution system. Those modes are called pressureless modes (see Sect. 2.7.4).

The influence of an axial velocity component in the core of a strained vortex was analyzed by Lacaze et al. [22]. They showed that the resonant Kelvin modes $m = 1$ and $m = -1$, which are the most unstable in the absence of axial flow, become damped as the axial flow is increased. This was shown to be due to the appearance of a critical layer which damps one of the resonant Kelvin modes. However, the elliptic instability did not disappear. Other combinations of Kelvin modes $m = -2$ and $m = 0$, then $m = -3$, and $m = -1$ were shown to become progressively unstable for increasing axial flow.

2.3.2 The Hyperbolic Instability

The hyperbolic instability is easier to understand for fluid without rotation and stratification. Then, when the strain, ϵ, is larger than the vorticity, ζ, the streamlines are hyperbolic as shown in Fig. 2.8 and the continuous stretching along the unstable manifold of the stagnation point of the flow induces instability. The instability

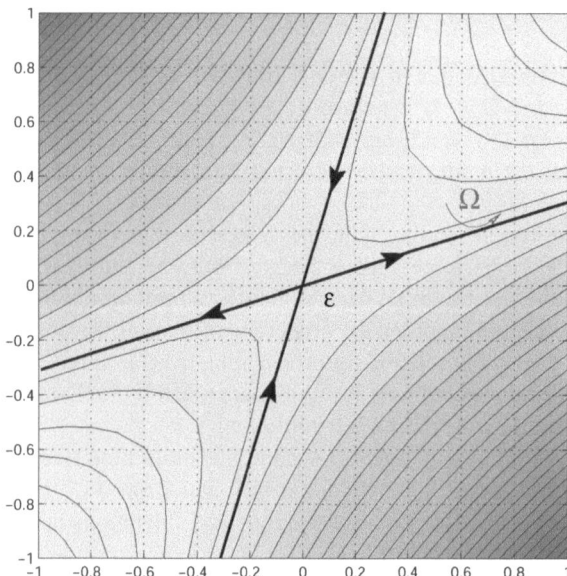

Fig. 2.8 Flow around an hyperbolic fixed point

modes have only vertical wave vectors and therefore the modes are "pressureless" since they are associated with zero pressure variations. This instability has been discussed in particular by Pedley [40], Caulfield and Kerswell [12], and Leblanc and Cambon [24]. Like for the previous case, no external length scales enter the problem and the hyperbolic instability affects the wave vectors independently of their modulus and is associated with a unique growth rate σ (see Sect. 2.7.2), including background rotation:

$$\sigma^2 = \epsilon^2 - (2\Omega + \zeta/2)^2. \qquad (2.19)$$

The stratification plays no role in the hyperbolic instability because the wave vector is vertical and thus the motion is purely horizontal. In the absence of background rotation, the hyperbolic instability develops only at hyperbolic points. In contrast, in the presence of an anticyclonic mean rotation, the hyperbolic instability can develop at elliptical points since σ may be real while $\zeta/2$ is larger than ϵ (see also Sect. 2.7.4).

2.4 The Zigzag Instability

All the previously discussed 3D instability mechanisms, except the 2D Kelvin–Helmholtz instability, are active at all vertical scales and preferentially at very small scales. Their growth rate scales like the inverse of the vortex turnover time. The last instability we would like to discuss has been introduced by Billant and Chomaz [4]. It selects a particular vertical wave number and has been proposed as the basic mechanism for energy transfer in strongly stratified turbulence. Thus we will first discuss the mechanism responsible for the zigzag instability in stratified flows in the absence of rotation. Next, rotation effects will be taken into account.

2.4.1 The Zigzag Instability in Strongly Stratified Flow Without Rotation

When the flow is strongly stratified the buoyancy length scale $L_B = U/N$ is assumed to be much smaller than the horizontal length scale L. In that case the vertical deformation of an iso-density surface is at most L_B^2/L_V (where L_V is the vertical scale) and therefore the velocity, which in the absence of diffusion should be tangent to the iso-density surface, is to leading order horizontal.

If we further assume, as did Riley et al. [46] and Lilly [31], that the vertical scale L_V is large compared to L_B, then the vertical stretching of the potential vorticity is negligible, since the vertical vorticity itself is (to leading order) conserved while being advected by the 2D horizontal flow. Similarly the variation of height of a column of fluid trapped between two iso-density surfaces separated by a distance

L_V is negligible, since the conservation of mass imposes to leading order that the horizontal velocity field is divergence free.

The motion is therefore governed to leading order by the 2D Euler equations independently in each layer of vertical size L_V as soon as $L_V \gg L_B$. To leading order, there is no coupling in the vertical. Having made this remark, Riley et al. [46] and Lilly [31] conjectured that the strongly stratified turbulence should be similar to the purely 2D turbulence and they invoked the inverse energy cascade of 2D turbulence to interpret measured velocity spectra in the atmosphere.

However, Billant and Chomaz [5] have shown that a generic instability is taking the flow away from the assumption $L_V \gg L_B$. The key idea is that there is no coupling across the vertical if the vertical scale of motion is large compared to the buoyancy length scale. Thus, we may apply to the vortex any small horizontal translations with a distance and possibly a direction that both vary vertically on a large scale compared to L_B. This means that, in the limit where the vertical Froude number $F_V = L_B/L_V = kL_B$ goes to zero, infinitesimal translations in any directions are neutral modes since they transform a solution of the leading order equation into another solution. Now if $F_V = kL_B$ is finite but small it is possible to compute the corrections and determine if the neutral mode at $kL_B = 0$ is the starting point of a stable or an unstable branch (see Billant and Chomaz [6]). Such modes are called phase modes since they are reminiscent of a broken continuous invariance (translation, rotation, etc.).

More precisely, in the case of two vortices of opposite sign, a detailed asymptotic analysis leads to two coupled linear evolution equations for the y position of the center of the dipole $\eta(z, t)$ and the angle of propagation $\phi(z, t)$ (see Fig. 2.9) up to fourth order in F_V:

$$\frac{\partial \eta}{\partial t} = \phi, \tag{2.20}$$

$$\frac{\partial \phi}{\partial t} = (D + F_h^2 g_1) F_V^2 \frac{\partial^2 \eta}{\partial z^2} + g_2 F_V^4 \frac{\partial^4 \eta}{\partial z^4}, \tag{2.21}$$

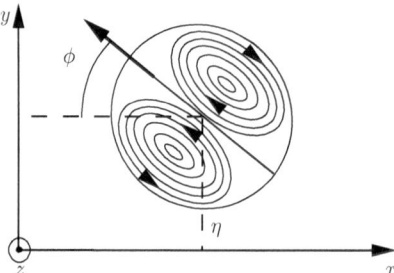

Fig. 2.9 Definition of the phase variables η and ϕ for the Lamb dipole, from Billant and Chomaz [6]

where $F_h = L_B/L$ is the horizontal Froude number and $D = -3.67$, $g_1 = -56.4$, and $g_2 = -16.1$. These phase equations show that when F_V is non-zero, the translational invariance in the direction perpendicular to the traveling direction of the dipole (corresponding to the phase variable η) is coupled to the rotational invariance (corresponding to ϕ). Substituting perturbations of the form $(\eta, \phi) = (\eta_0, \phi_0) \exp(\sigma t + ikz)$ yields the dispersion relation

$$\sigma^2 = -(D + g_1 F_h^2) F_V^2 k^2 + g_2 F_V^4 k^4. \tag{2.22}$$

Perturbations with a sufficiently long wavelength ($F_V \ll 1$) are always unstable because the coefficients D and g_1 are negative. There is, however, a stabilization at large wavenumbers since g_2 is negative. Therefore, because the similarity parameter in (2.22) is $k F_V$, the selected wavelength will scale like L_B whereas the growth rate will stay constant and scale like U/L. This instability therefore invalidates the initial assumption that the vertical length scale is large compared to the buoyancy length scale. Similar phase equations have been obtained for two co-rotating vortices [39]. In this case, the rotational invariance is coupled to an invariance derived from the existence of a parameter describing the family of basic flows: the separation distance b between the two vortex centers. This leads to two phase equations for the angle of the vortex pair $\alpha(z, t)$ and for $\delta b(z, t)$ the perturbation of the distance separating the two vortices:

$$\frac{\partial \alpha}{\partial t} = -\frac{2\Gamma}{\pi b^3} \delta b + \frac{\Gamma}{\pi b} D_0 F_V^2 \frac{\partial^2 \delta b}{\partial z^2}, \tag{2.23}$$

$$\frac{\partial \delta b}{\partial t} = -\frac{\Gamma b}{\pi} D_0 F_V^2 \frac{\partial^2 \alpha}{\partial z^2}, \tag{2.24}$$

where Γ is the vortex circulation and $D_0 = (7/8) \ln 2 - (9/16) \ln 3$ is a coefficient computed from the asymptotics. The dispersion relation is then

$$\sigma^2 = -\frac{\Gamma^2}{\pi^2} \left(\frac{2}{b^2} D_0 (F_V k)^2 + D_0^2 (F_V k)^4 \right). \tag{2.25}$$

There is a zigzag instability for long wavelengths because D_0 is negative. This theoretical dispersion relation is similar to the previous one for a counter-rotating vortex pair except that the most amplified wavenumber depends not only on F_V but also on the separation distance b. This is in very good agreement with results from numerical stability analyses [37].

For an axisymmetric columnar vortex, the phase mode corresponds to the azimuthal wave number $m = 1$, and at $kL_B = 0$ the phase mode is associated to a zero frequency. In stratified flows, as soon as a vortex is not isolated, this phase mode may be destabilized by the strain due to other vortices.

2.4.2 The Zigzag Instability in Strongly Stratified Flow with Rotation

If the fluid is rotating, Otheguy et al. [38] have shown that the zigzag instability continues to be active with a growth rate almost constant (Fig. 2.10). However, the wavelength varies with the planetary rotation Ω and scales like $|\Omega| L / N$ for small Rossby number in agreement with the quasi-geostrophic theory. The zigzag instability then shows that quasi-geostrophic vortices cannot be too tall as previously demonstrated by Dritschel and de la Torre Juárez [16].

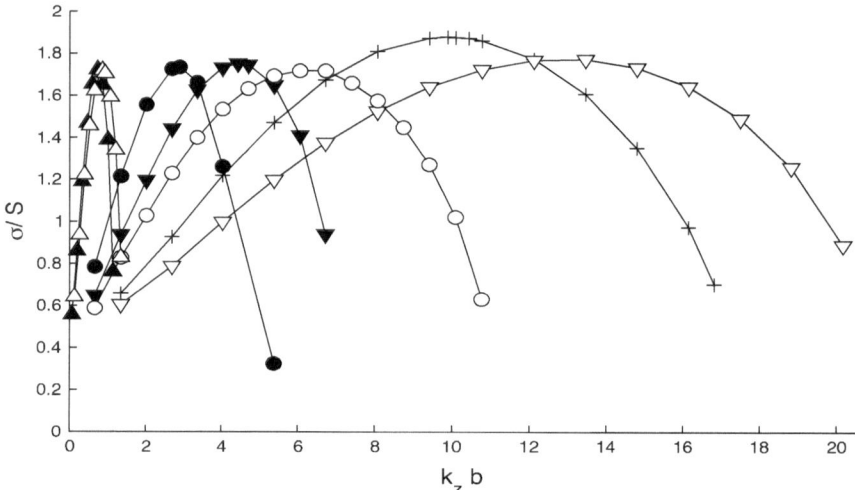

Fig. 2.10 Growth rate of the zigzag instability normalized by the strain rate $S = \Gamma/(2\pi b^2)$ plotted against the vertical wavenumber k_z scaled by the separation distance b for $F_h = \Gamma/(2\pi R^2 N) = 0.5$ (R is the vortex radius), $Re = \Gamma/(2\pi \nu) = 8000$, $R/b = 0.15$ and for $Ro = \Gamma/(2\pi R^2 \Omega) = \infty$ (+), $Ro = \pm 2.5$ ($\triangledown$), $Ro = \pm 1.25$ ($\circ$), $Ro = \pm 0.25$ ($\triangle$). Cyclonic rotations are represented by *filled symbols* whereas anticyclonic rotations are represented by *open symbols*. From [38]

2.5 Experiment on the Stability of a Columnar Dipole in a Rotating and Stratified Fluid

This last section presents results of an experiment on a vortex pair in a rotating and stratified fluid [14, 8] that illustrates many of the instabilities previously discussed that tends to induce 3D motions.

2.5.1 Experimental Setup

As in Billant and Chomaz [4] a tall vertical dipole is created by closing a double flap apparatus as one would close an open book (Fig. 2.11). This produces a dipole

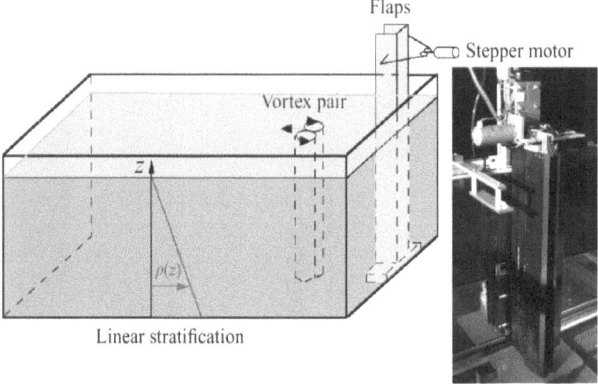

Fig. 2.11 Sketch of the experimental setup that was installed on the rotating table of the Centre National de Recherches Météorologiques (Toulouse). The flaps are 1 m tall and the tank is 1.4 m long, and 1.4 m large, 1.4 m deep [14, 8]

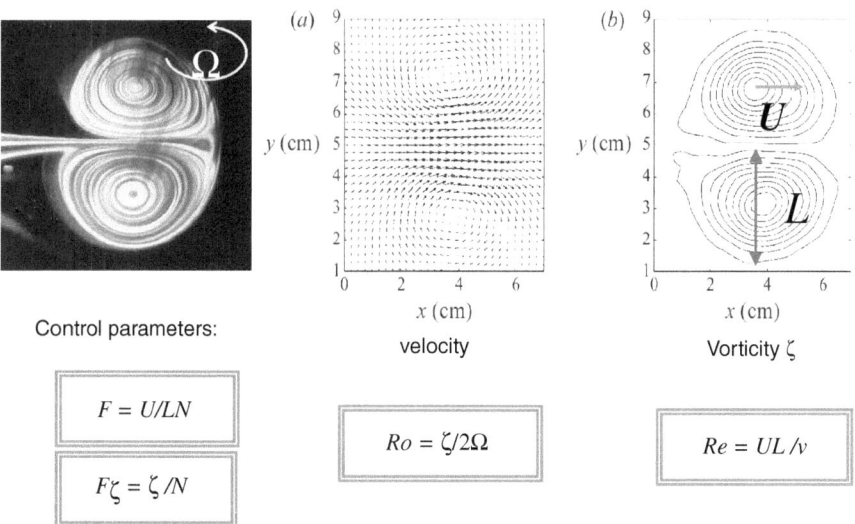

Fig. 2.12 Flow parameters for a dipole in a stratified or rotating fluid

that moves away from the flaps and, in the absence of instability, is vertical. Particle image velocimetry (PIV) measurements provide the dipole characteristics that are used to compute the various parameters (Fig. 2.12).

2.5.2 The State Diagram

Depending upon the value of the Rossby number Ro and Froude number F_ζ, the different types of instabilities described in the previous sections are observed (Fig. 2.13). Positive Rossby numbers correspond to instabilities observed on cyclonic

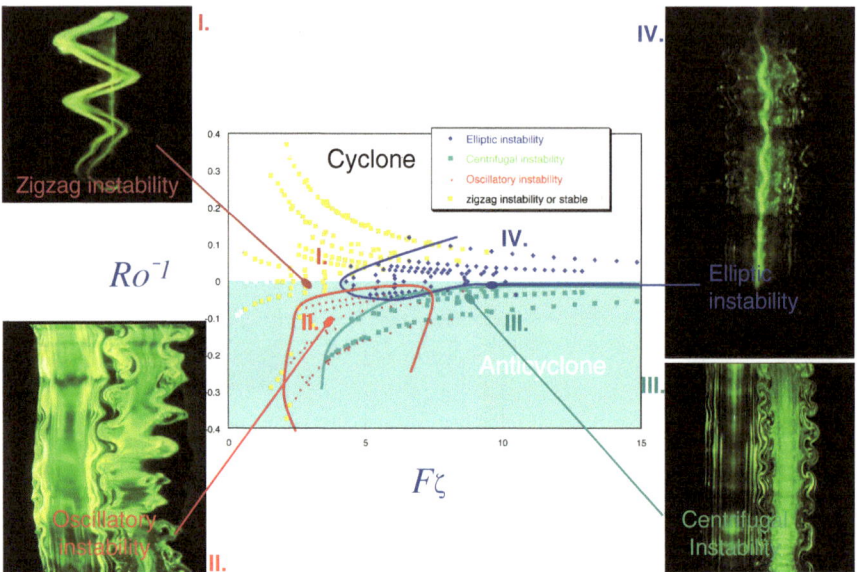

Fig. 2.13 The state diagram where the different instabilities are observed [14, 8] (see also Sect. 2.1 for more details on centrifugal instability)

vortices, while negative Rossby numbers correspond to instabilities observed on anticyclones. For large Rossby number, Colette et al. [14] and Billant et al. [8] have observed the zigzag instability at small Froude number and the elliptic instability at large Froude number on both vortices as in Billant and Chomaz [4]. As the Rossby number is decreased, the elliptic instability develops with different wavelengths and growth rates on the cyclone and the anticyclone. For smaller Rossby number, the elliptic instability continues to be observed on the cyclone but tends to be stabilized by rotation effects beyond a given Froude number. In contrast, the anticyclone becomes subjected to two other types of instability: a centrifugal instability for large Froude number and an oscillatory asymmetric instability for moderate Froude number.

2.6 Discussion: Instabilities and Turbulence

Experimental results as well as theoretical results show that when the strain field is large enough, a quasi-two-dimensional vortex is never stable versus 3D instabilities. Regarding the elliptic, hyperbolic, and centrifugal instabilities, if the strain is small, only anticyclones are stable in a narrow band between $\max\left[-1, -(1/2 + \epsilon/\zeta)^{-1}\right] < Ro < -2/3$ if $F > 2$ and if $F < 2$, vortices are stable for $Ro > -2/3$. All the instabilities which have been described have a growth rate scaling like the vorticity magnitude or strain field induced by the other vortices. This means that these instabilities are as fast as the mechanisms usually invoked for the turbulence cascade, such as the pairing of same sign vortices or the creation of vorticity filaments. They should therefore modify the phenomenology of the turbulence. In particular for large Rossby number and small Froude number, the pairing of two vortices is unstable to the generalized zigzag instability and, while approaching each other the vortices should form thinner and thinner layers resulting in an energy transfer to smaller length scales and not larger scales as in the 2D turbulence. This effect of the zigzag instability would then explain the result of Lindborg [32] who shows, processing turbulence data collected during airplane flights and computing third-order moment of the turbulence, that the energy cascade in the horizontal energy spectra is direct and not reverse as it was conjectured by Lilly [31].

2.7 Appendix: Local Approach Along Trajectories

In this appendix, we investigate the instability of a 2D steady basic flow characterized by a velocity $\mathbf{U}_B$ and a pressure P_B in the inviscid case. The normal vector of the flow field is denoted $\mathbf{e}_z$. The 3D perturbations are denoted $(\mathbf{u}, p)$. In a frame, rotating at the angular frequency $\mathbf{\Omega} = \Omega \mathbf{e}_z$, the linearized Euler equation read as

$$\frac{\partial \mathbf{u}}{\partial t} + \mathbf{u}.\nabla \mathbf{U}_B + \mathbf{U}_B \nabla \mathbf{u} = -\nabla p - 2\mathbf{\Omega} \times \mathbf{u}, \qquad (2.26)$$

$$\nabla.\mathbf{u} = 0. \qquad (2.27)$$

Following Lifschitz and Hameiri [33], we consider a rapidly oscillating localized perturbation:

$$\mathbf{u} = \exp\{i\phi(\mathbf{x}, t)/\xi\}\, \mathbf{a}(t) + O(\delta), \qquad (2.28)$$

$$p = \exp\{i\phi(\mathbf{x}, t)/\xi\}\,(\pi(t) + O(\delta)), \qquad (2.29)$$

where ξ is a small parameter. Substitution in the linearized Euler equations leads to a set of differential equations for the amplitude and the wave vector $\mathbf{k} = \nabla\phi$ evolving along the trajectories of the basic flow:

$$\frac{d\mathbf{x}}{dt} = \mathbf{U}_B, \qquad (2.30)$$

$$\frac{d\mathbf{k}}{dt} = -\mathcal{L}_B^T \mathbf{k}, \qquad (2.31)$$

$$\frac{d\mathbf{a}}{dt} = \left(\frac{2\mathbf{k}\mathbf{k}^T}{|\mathbf{k}|^2} - \mathcal{I}\right)\mathcal{L}_B \mathbf{a} + \left(\frac{\mathbf{k}\mathbf{k}^T}{|\mathbf{k}|^2} - \mathcal{I}\right)\mathcal{C}\mathbf{a}, \qquad (2.32)$$

with $\mathcal{L}_B = \nabla\mathbf{U}_B$ the velocity gradient tensor, the superscript T denoting the transposition, $\mathcal{I}$ the unity tensor, and $\mathcal{C}$ the Coriolis tensor:

$$\mathcal{C} = \begin{pmatrix} 0 & -2\Omega & 0 \\ 2\Omega & 0 & 0 \\ 0 & 0 & 0 \end{pmatrix}. \tag{2.33}$$

The pressure has been eliminated by applying the operator:

$$\mathcal{P}(\mathbf{k}) = \left(\mathcal{I} - \frac{\mathbf{k}\mathbf{k}^T}{|\mathbf{k}|^2} \right). \tag{2.34}$$

The incompressibility equation (2.27) yields

$$\mathbf{k}.\mathbf{a} = 0. \tag{2.35}$$

The stability is analyzed looking for the behavior of the velocity amplitude $\mathbf{a}$. Following Lifschitz and Hameiri [33], the flow is unstable if there exists a trajectory on which the amplitude $\mathbf{a}$ is unbounded at large time.

2.7.1 Centrifugal Instability

For flows with closed streamlines, centrifugal instability can be understood by considering spanwise perturbations, as they have been shown to be the most unstable in centrifugal instability studies (see [2] or [48] for a detailed discussion). If $\mathbf{k}(t = 0) = k_0 \mathbf{e_z}$, (2.31) yields $\mathbf{k}(t) = k_0 \mathbf{e_z}$; the base flow, evolving in a plane perpendicular to $\mathbf{e_z}$, does not impose tilting or stretching along z direction. The incompressibility equation (2.35) gives $a_z = 0$.

In the plane of the steady base flow, the trajectories (streamlines in the steady case) may be referred to as streamline function value ψ. The two vectors, $\mathbf{U}_B$ and $\nabla \psi$, provide an orthogonal basis. The amplitude equation in the plane perpendicular to $\mathbf{e_z}$ may be expressed in this new coordinate system, and following [48], (2.32) becomes

$$\frac{d}{dt} \begin{pmatrix} \mathbf{a}.\mathbf{U}_B \\ \mathbf{a}.\nabla\psi \end{pmatrix} = \begin{pmatrix} 0 & \zeta + 2\Omega \\ -2(\frac{V}{\mathcal{R}} + \Omega) & 2V'/V \end{pmatrix} \begin{pmatrix} \mathbf{a}.\mathbf{U}_B \\ \mathbf{a}.\nabla\psi \end{pmatrix}, \tag{2.36}$$

with $V = |\mathbf{U}_B(\mathbf{x})|$, V' the lagrangian derivative $\frac{d}{dt}V = \mathbf{U}_B.\nabla V$ and $\mathcal{R}$ the local algebraic curvature radius defined by [48]

$$\mathcal{R}(x, y) = \frac{V^3}{\nabla\psi.(\mathbf{U}_B.\nabla\mathbf{U}_B)}. \tag{2.37}$$

Note that the generalized Rayleigh discriminant, $\delta = 2\left(\frac{V}{\mathcal{R}} + \Omega\right)(\zeta + 2\Omega)$, appears in (2.36) as the opposite of the determinant of the governing matrix. Sipp and Jacquin [48] showed that if there exists a particular streamline on which $\delta < 0$, the velocity amplitude, $\mathbf{a}$, is unbounded at large time. However, the general proof for the existence of diverging solutions of (2.36) for a closed (non-circular) streamline with δ negative requires further mathematical analyses.

2.7.2 Hyperbolic Instability

In this section, we focus again on a spanwise perturbation characterized by a wavenumber $\mathbf{k}$ perpendicular to the flow field. The contribution of the pressure disappears in that case and the instability is called pressureless. Equation (2.32) reduces to

$$\frac{d\mathbf{a}}{dt} = -\mathcal{L}_B \mathbf{a} - \mathcal{C}\mathbf{a}. \tag{2.38}$$

We look for instability of a basic flow characterized by

$$\mathcal{L}_B = \begin{pmatrix} 0 & \epsilon - \frac{\zeta}{2} & 0 \\ \epsilon + \frac{\zeta}{2} & 0 & 0 \\ 0 & 0 & 0 \end{pmatrix}, \tag{2.39}$$

with ϵ the local strain and ζ the vorticity of the base flow. For $|\epsilon|$ greater than $|\frac{\zeta}{2}|$ the streamlines are hyperbolic. The integration of (2.38) is straightforward. The perturbed velocity amplitude $\mathbf{a}$ has an exponential behavior, $\exp(\sigma t)$, with σ

$$\sigma = \sqrt{\epsilon^2 - (\zeta/2 + 2\Omega)^2}, \tag{2.40}$$

as discussed in Sect. 2.3.2.

2.7.3 Elliptic Instability

The basic flow is defined by the constant gradient velocity tensor (2.39) but with $|\frac{\zeta}{2}|$ greater than $|\epsilon|$. Trajectories are found to be elliptical with an aspect ratio or eccentricity:

$$E = \frac{\sqrt{\frac{\zeta}{2} - \epsilon}}{\sqrt{\frac{\zeta}{2} + \epsilon}}. \tag{2.41}$$

Along those closed trajectories, (2.31) has period

$$T = \frac{2\pi}{Q} = \frac{2\pi}{\sqrt{\left(\frac{\zeta}{2}\right)^2 - \epsilon^2}}. \tag{2.42}$$

The wave vector solution of (2.31),

$$\mathbf{k} = k_0 \left(\sin(\theta) \cos(Q(t - t_0)), \; E \; \sin(\theta) \sin(Q(t - t_0)), \; \cos(\theta) \right), \quad (2.43)$$

with k_0 and t_0 integration constants, is easily obtained (see, for example, Bayly [3]). The wave vector describes an ellipse parallel to the (x, y) plane, with the same eccentricity as the trajectories ellipse, but with the major and minor axes reversed. The stability is investigated with Floquet theory, looking for the eigenvalues of the monodromy matrix $\mathcal{M}(T)$, which is a solution of

$$\frac{d\mathcal{M}}{dt} = \left(\frac{2\mathbf{kk}^T}{|\mathbf{k}|^2} - \mathcal{I} \right) L_B \mathcal{M} + \left(\frac{\mathbf{kk}^T}{|\mathbf{k}|^2} - \mathcal{I} \right) C \mathcal{M}, \quad (2.44)$$

$$\text{with} \quad \mathcal{M}(0) = \mathcal{I}, \quad (2.45)$$

integrated over one period T. The basic flow lies in the (x, y) plane, so that one of the eigenvalues of the monodromy matrix is $m_3 = 1$. The average of the trace of the matrix $\left(\frac{2\mathbf{kk}^T}{|\mathbf{k}|^2} - \mathcal{I} \right) L_B + \left(\frac{\mathbf{kk}^T}{|\mathbf{k}|^2} - \mathcal{I} \right) C$ is zero, making the determinant of $\mathcal{M}(T)$ equal to unity. The two other eigenvalues of $\mathcal{M}(T)$ are then $m_1 m_2 = 1$, with $m_{1,2}$ either complex conjugate indicating stable flow, or real and inverse for unstable flow. Generally, $m_{1,2}$ are obtained by numerical integration of (2.44) and (2.45). Note, however, that the following two cases may be tackled analytically.

2.7.4 Pressureless Instability

In the case of a pressureless instability the monodromy equation is a solution of

$$\frac{d\mathcal{M}}{dt} = -L_B \mathcal{M} - C \mathcal{M} \quad (2.46)$$

$$\mathcal{M}(0) = \mathcal{I}. \quad (2.47)$$

The eigenvalues of the monodromy equation are

$$m_i = \exp\left(\pm \sqrt{\epsilon^2 - (\zeta/2 + 2\Omega)^2} \, T \right), \quad \text{with } i = 1, 2. \quad (2.48)$$

With rotation, the elliptical flow may be unstable to pressureless instabilities, as discussed in Sects. 2.3.1 and 2.3.2 and in Craik [15].

2.7.5 Small Strain $|\epsilon \ll 1|$

Following Waleffe [50], we derive from (2.32) the equation for the rescaled component of the velocity along the z-axis denoted q:

$$q = a_3 \frac{|\mathbf{k}|^2}{|\mathbf{k}_{//}|^2}, \tag{2.49}$$

with $\mathbf{k}_{//}$ the component of the wavenumber on the (x, y) plane. For small strain, this equation leads to a Mathieu equation with a rescaled time, $t^\star = \zeta t$,

$$\frac{d^2 q}{dt^{\star 2}} + \left(\alpha + 2\epsilon b \sin t^\star \right) q = 0, \tag{2.50}$$

where

$$\alpha = \left(\frac{Ro + 1}{Ro} \right)^2 \cos^2 \theta, \tag{2.51}$$

$Ro = \frac{\zeta}{2\Omega}$ is the Rossby number and

$$b = \frac{1}{2} \left[\left(\frac{Ro + 1}{Ro} \right)^2 \sin^2 \theta + \frac{1}{Ro} + \frac{3}{2} \right] \cos^2 \theta. \tag{2.52}$$

Parametric resonances occur when

$$\alpha = \frac{1}{4} j^2, \tag{2.53}$$

with j an integer, giving the formula (2.14) discussed in Sect. 2.3.1.

References

1. Armi, L., Hebert, D., Oakey, N., Price, J., Richardson, P.L., Rossby, T., Ruddick, B.: The history and decay of a Mediterranean salt lens. Nature, **333**, 649 (1988).
2. Bayly, B.J.: Three-dimensional centrifugal-type instabilities in inviscid two-dimensional flows. Phys. Fluids **31**, 56 (1988).
3. Bayly, B.J.: Three-dimensional instability of elliptical flow. Phys. Rev. Lett. **57**, 2160 (1986).
4. Billant, P., Chomaz, J.-M.: Experimental evidence for a new instability of a vertical columnar vortex pair in a strongly stratified fluid. J. Fluid Mech. **418**, 167 (2000).
5. Billant, P., Chomaz, J.-M.: Theoretical analysis of the zigzag instability of a vertical columnar vortex pair in a strongly stratified fluid. J. Fluid Mech. **419**, 29 (2000).
6. Billant, P., Chomaz, J.-M.: Three-dimensional stability of a vertical columnar vortex pair in a stratified fluid. J. Fluid Mech. **419**, 65 (2000).
7. Billant, P., Colette, A., Chomaz, J.-M.: Instabilities of anticyclonic vortices in a stratified rotating fluid. Bull. Am. Phys. Soc. **47**(10), 162. In: Proceedings of the 55th Meeting of the APS Division of Fluid Dynamics, Dallas, 24–26 Nov 2002.
8. Billant, P., Colette, A., Chomaz, J.-M.: Instabilities of a vortex pair in a stratified and rotating fluid. In: Proceedings of the 21st International Congress of the International Union of Theoretical and Applied Mechanics, Varsovie, 16–20 Aug 2004.
9. Billant, P., Gallaire, F.: Generalized Rayleigh criterion for non-axisymmetric centrifugal instabilities. J. Fluid Mech. **542**, 365–379 (2005).

10. Cambon, C.: Turbulence and vortex structures in rotating and stratified flows. Eur. J. Mech.
    B Fluids **20**, 489 (2001).
11. Carton, X., McWilliams, J.: Barotropic and baroclinic instabilities of axisymmetric vortices
    in a quasi-geostrophic model. In: Nihoul, J., Jamart, B. (eds.) Mesoscale/Synoptic Coherent
    Structures in Geophysical Turbulence, p. 225. Elsevier, Amsterdam (1989).
12. Caulfield, C.P., Kerswell, R.R.: The nonlinear development of three-dimensional disturbances
    at hyperbolic stagnation points: a model of the braid region in mixing layers. Phys. Fluids **12**,
    1032 (2000).
13. Chomaz, J.-M., Rabaud, M., Basdevant, C., Couder, Y.: Experimental and numerical investi-
    gation of a forced circular shear layer. J. Fluid Mech. **187**, 115 (1988).
14. Colette, A., Billant, P., Beaudoin, B., Boulay, J.C., Lassus-Pigat, S., Niclot, C., Schaffner,
    H., Chomaz, J.-M.: Instabilities of a vertical columnar vortex pair in a strongly stratified and
    rotating fluid. Bull. Am. Phys. Soc. **45**(9), 175. In: Proceeding of the 53rd Meeting of the
    APS Division of Fluid Dynamics, Washington, 19–21 Nov 2000.
15. Craik, A.D.D.: The stability of unbounded two- and three-dimensional flows subject to body
    forces: some exact solutions J. Fluid Mech. **198**, 275 (1989).
16. Dritschel, D.G., Torre Juárez, M.: The instability and breakdown of tall columnar vortices in
    a quasi-geostrophic fluid. J. Fluid Mech. **328**, 129 (1996).
17. Eckhoff, K.S.: A note on the instability of columnar vortices. J. Fluid Mech. **145**, 417 (1984).
18. Eckhoff, K.S., Storesletten, L.: A note on the stability of steady inviscid helical gas flows.
    J. Fluid Mech. **89**, 401 (1978).
19. Fontane, J.: Etude expérimentale des instabilités 3-D d'une paire de tourbillons en milieu
    tournant. Master thesis, CNRM, LadHyX, École Polytechnique. DEA Dynamique des Flu-
    ides, Toulouse. (2002).
20. Kerswell, R.: Elliptical instability. Annu. Rev. Fluid Mech. **34**, 83 (2002).
21. Kloosterziel, R., Van Heijst, G.: An experimental study of unstable barotropic vortices in a
    rotating fluid. J. Fluid. Mech. **223**, 1 (1991).
22. Lacaze, L., Ryan, K., Le Dizès, S.: Elliptic instability in a strained Batchelor vortex. J. Fluid.
    Mech. **577**, 341–361 (2007).
23. Leblanc, S.: Internal wave resonances in strain flows. J. Fluid. Mech. **477**, 259 (2003).
24. Leblanc, S., Cambon, C.: On the three-dimensional instabilities of plane flows subjected to
    Coriolis force. Phys. Fluids **9**, 1307 (1997).
25. Leblanc, S., Le Duc, A.: The unstable spectrum of swirling gas flows. J. Fluid Mech. **537**,
    433–442 (2005).
26. Le Dizès, S.: Inviscid waves on a Lamb–Oseen vortex in a rotating stratified fluid: conse-
    quences for the elliptic instability. J. Fluid Mech. **597**, 283 (2008).
27. Le Dizès, S., Eloy, C.: Short-wavelength instability of a vortex in a multipolar strain field.
    Phys. Fluids **11**, 500 (1999).
28. Leibovich, S., Stewartson, K.: A sufficient condition for the instability of columnar vortices.
    J. Fluid Mech. **126**, 335–356 (1983).
29. Ludwieg, H.: Stabilität der Strömung in einem zylindrischen Ringraum. Z. Flugwiss. **8**,
    135–142 (1960).
30. Eloy, C., Le Dizès, S.: Three-dimensional instability of Burgers and Lamb–Oseen vortices in
    a strain field. J. Fluid Mech. **378**, 145 (1999).
31. Lilly, D.K.: Stratified turbulence and the mesoscale variability of the atmosphere. J. Atmos.
    Sci. **40**, 749 (1983).
32. Lindborg, E.: Can the atmosphere kinetic energy spectrum be explained by two-dimensional
    turbulence? J. Fluid Mech. **388**, 259 (1999).
33. Lifschitz, A., Hameiri, E.: Local stability conditions in fluid dynamics. Phys. Fluids. **3**, 2644
    (1991).
34. Miyazaki, T.: Elliptical instability in a stably stratified rotating fluid. Phys. Fluids **5**, 2702
    (1993).
35. Moore, D.W., Saffman, P.G.: The instability of a straight vortex filament in a strain field. Proc.
    R. Soc. Lond. A **346**, 413 (1975).

36. Orlandi, P., Carnevale, G.: Evolution of isolated vortices in a rotating fluid of finite depth. J. Fluid Mech. **381**, 239 (1999).
37. Otheguy, P., Chomaz, J.M., Billant, P.: Elliptic and zigzag instabilities on co-rotating vertical vortices in a stratified fluid. J. Fluid Mech. **553**, 253 (2006).
38. Otheguy, P., Billant, P., Chomaz, J.-M.: The effect of planetary rotation on the zigzag instability of co-rotating vortices in a stratified fluid. J. Fluid Mech. **553**, 273 (2006).
39. Otheguy, P., Billant, P., Chomaz, J.-M.: Theoretical analysis of the zigzag instability of a vertical co-rotating vortex pair in a strongly stratified fluid. J. Fluid Mech. **584**, 103 (2007).
40. Pedley, T.: On the instability of viscous flow in a rapidly rotating pipe. J. Fluid Mech. **35**, 97 (1969).
41. Pierrehumbert, R.T.: Universal short-wave instability of two-dimensional eddies in an inviscid fluid. Phys. Rev. Lett. **57**, 2157 (1986).
42. Rabaud, M., Couder, Y.: A shear-flow instability in a circular geometry. J. Fluid. Mech. **136**, 291 (1983).
43. Rayleigh, L.: On the stability, or instability of certain fluid motions. Proc. Lond. Math. Soc. **11**, 57 (1880).
44. Rayleigh, L.: On the instability of cylindrical fluid surfaces. Phil. Mag. **34**, 177 (1892).
45. Rayleigh, L.: On the dynamics of revolving fluids. Proc. R. Soc. Lond. A **93**, 148 (1916).
46. Riley, J.J., Metcalfe, R.W., Weissman, M.A.: Direct numerical simulations of homogeneous turbulence in density stratified fluids. In: West, B.J. (ed.) Proceedings of the AIP Conference on Nonlinear Properties of Internal Waves, La Jolla, p. 79 (1981).
47. Synge, J.L.: The stability of heterogeneous liquids. Trans. R. Soc. Can. **27**(3), 1 (1933).
48. Sipp, D., Jacquin, L.: Three-dimensional centrifugal-type instabilities of two-dimensional flows in rotating systems. Phys. Fluids **12**, 1740 (2000).
49. Tsai, C.Y., Widnall, S.E.: The stability of short waves on a straight vortex filament in a weak externally imposed strain field. J. Fluid Mech. **73**, 721–733 (1976).
50. Waleffe, F.: On the three-dimensional instability of strained vortices. Phys. Fluids A **2**, 76 (1990).

Chapter 3
Oceanic Vortices

X. Carton

Oceanic vortices (also called eddies) come under a large variety of sizes, from a few kilometers up to 300 km in diameter. Vertically, their extent can also range from a few tens of meters up to (nearly) the whole ocean depth. They can be intensified near the surface, near the thermocline, or near the bottom. They can be generated by the instability or by the change of direction of ocean currents, by geostrophic adjustment after convection, or via topographic influences (e.g., lee eddies behind islands). But nearly all oceanic vortices share the common properties, which may be used to define them:

Oceanic vortices are coherent structures, with a dominant horizontal motion and closed fluid circulation in their core. The predominance of horizontal motion is due to the importance of the Coriolis force and of the buoyancy effects in the dynamics of vortices and to their small aspect ratio. Vortex lifetimes are usually much longer than the timescale of their spinning motion. In general, mixing and ventilation affect oceanic vortices on timescales much larger than the turnover time. But such mixing processes can also have drastic consequences when they reach the vortex core: then the vortex collapses. Since mixing is generally slow and since their flow pattern is quasi-circular, oceanic vortices retain in their core a water mass characteristic of their region of formation. Thus, oceanic vortices, which drift over long distances, participate in the transport of heat, momentum, chemical tracers, and biological species across ocean basins and contribute to the mixing of oceanic water masses.

The present review will first concentrate on oceanic observations of vortices, on the description of their physical characteristics, and on the salient features of their dynamics. The second part of this review will present dynamical models often used to represent vortex motion, evolution, and interactions and recent findings on these subjects.

X. Carton (✉)
Laboratoire de Physique des Oceans, Universite de Bretagne Occidentale, Brest, France,
xcarton@univ-brest.fr

Carton, X.: *Oceanic Vortices*. Lect. Notes Phys. **805**, 61–108 (2010)
DOI 10.1007/978-3-642-11587-5_3 ⓒ Springer-Verlag Berlin Heidelberg 2010

3.1 Observations of Oceanic Vortices

3.1.1 Different Types of Oceanic Vortices

First, we will present the large, surface-intensified "rings" of the major western boundary currents which have been historically identified and studied first; then, we will describe smaller (mesoscale) vortices, identified later on the eastern boundary of the oceans, but of importance for the large-scale fluxes of heat and salt. Some of these mesoscale eddies are concentrated at depth (for instance, in the thermocline) and thus their identification and study have been more recent.

3.1.1.1 Large Rings

In general, wind-induced currents are intensified at the western boundaries of the ocean and detach at mid-latitudes to form intense, horizontally and vertically sheared, eastward jets, prone to barotropic and baroclinic instabilities. These instabilities cause these jets to meander, the occlusion of the meanders resulting in the formation of so-called rings or synoptic eddies. These rings, and in particular the warm-core rings which are surface intensified, have long since been identified in the vicinity of the Gulf Stream, of the Kuroshio, of the Agulhas Current, or of the North Brazil Current, to name a few [140, 76]. Rings were so called because the original current circles on itself, so that the velocity maximum is then located on a ring that encircles and isolates a core with a trapped water mass.

Gulf Stream Rings

The Gulf Stream is the fastest current in the North Atlantic Ocean; it detaches from the American coast at Cape Hatteras to enter the Atlantic basin as an intense, quasi-zonal jet. Its peak velocity is on the order of 1.5 m/s, the jet width is about 80 km, and it is intensified above the main thermocline (roughly the upper 800 m of the ocean); below, the jet velocity is usually less than 0.1 m/s; Fig. 3.2 also shows that the isotherms dive by 600 m across the Gulf Stream (see also [131]). As a strongly sheared current, the Gulf Stream is unstable and forms meanders which can grow, occlude, and detach from the jet, forming anticyclonic/cyclonic rings on its northern/southern flanks ([166]; see Fig. 3.1). Since the Gulf Stream separates the warm waters of the Sargasso Sea from the cold waters of the Blake Plateau (see Fig. 3.2), cyclonic rings carry these cold waters and anticyclonic rings the warm waters. Cold-core rings are usually wider than warm-core rings (250 versus 150 km in diameter, [117]), and they extend down to 4000 m depth whereas warm-core rings are concentrated above the thermocline [137]. The maximum orbital velocity of the rings is comparable to the peak velocity of the Gulf Stream (about 1 or 1.5 m/s); it lies at a 30–40 km distance from the vortex center and decreases exponentially beyond [116]. In warm-core rings, intense velocities are still found at mid-depth, as in the Gulf Stream itself (e.g., 0.5 m/s at 500 m depth, [131]).

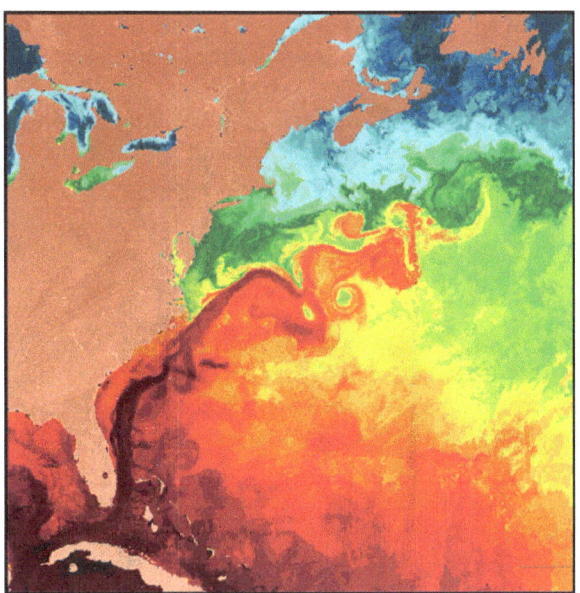

Fig. 3.1 Sea surface temperature in the North Atlantic Ocean in June 1984 showing the Gulf Stream separating the colder waters (in *green* and *blue*) of the North Mid-Atlantic Bight and Georges Bank, north, from the warmer waters (in *red*) of the Sargasso Sea south. Note the long meanders on the jet and the separated rings (image from the Coastal Carolina University web site, http://kingfish.coastal.edu/marine/gulfstream, courtesy Craig Gilman, CCU)

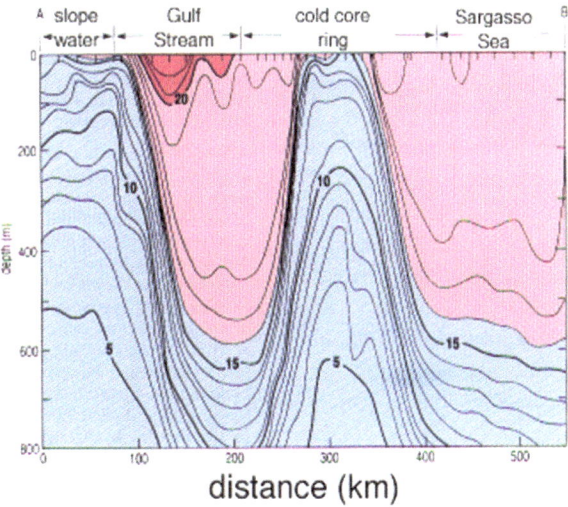

Fig. 3.2 Vertical section of temperature across the Gulf Stream and one of its cold-core rings showing the cold waters of the North Mid-Atlantic Bight and Georges Bank (in *blue*) and the warm waters of the Sargasso Sea south (in *pink*; image from the Coastal Carolina University web site, http://kingfish.coastal.edu/marine/gulfstream, courtesy Craig Gilman, CCU)

About 10–15 rings (both parities added) are generated every year; a few are rapidly re-absorbed by the Gulf Stream, while others may live much longer: warm-core rings undergo a destructive influence of topography north of the jet, combined with atmospheric ventilation, so that their lifetime does not generally exceed a few months. On the contrary, cold-core rings can drift across the Sargasso Sea for up to 3 years. Some of them end up near Cape Hatteras where they can interact with their parent jet and induce a meridional oscillation of the jet axis.

Agulhas Rings

Another intense western boundary current producing rings is the Agulhas Current. The Agulhas Current is fast (2 m/s near the surface), narrow (about 80–100 km), and concentrated in the upper 1000 m of the ocean ([90] and references therein). Its transport (about 95 Sv, with $1\,\text{Sv} = 10^6\,\text{m}^3/\text{s}$) is comparable to that of the Gulf Stream (100–130 Sv).

South of Africa, the Agulhas Current veers on itself (it turns anticlockwise to head east as the Agulhas Return Current, parallel to, and north of the Antarctic Circumpolar Current). This change of direction has been widely studied and is known as retroflection [92]; its origin has been attributed to potential vorticity conservation. As the current heads south, it gains planetary vorticity and must acquire anticyclonic relative vorticity [47]. Therefore, it must veer counterclockwise. Another possibility for the current to conserve potential vorticity is to increase its depth as it goes poleward; but this depth increase is physically limited so that the current must finally stabilize zonally at a given latitude [63, 91]. Other mechanisms proposed to explain retroflection implied the wind stress curl distribution [46], which vanishes south of the African continent or bottom topography.

The Agulhas rings are generated at the retroflection of the Agulhas Current (contrary to the Gulf Stream which produces rings essentially far from the coast). To form rings, the retroflecting current extends westward and then recedes eastward, thus isolating a loop which closes on itself (often near 16–18E, 38–40S). This zonal pulsation of the Agulhas Current has been attributed to variations in the current itself (variations in transport or advection of solitary meanders called Natal Pulses) or to variations in its surroundings (presence of nearby eddies, in particular cyclonic, or local current shears). This ring generation mechanism has long since been studied [18, 17, 32]. Periodic eddy shedding ensuring momentum conservation [114, 127] or barotropic instability of the Agulhas Current [48] has been identified as ring generation mechanisms.

The Agulhas rings are anticyclonic, with azimuthal velocities of 0.5–1 m/s, have diameters ranging from 250 to 400 km, and are often accompanied by smaller cyclones. These rings are intensified above 1000 m depth, but their dynamical extent can reach the bottom of the ocean, due to their barotropic velocity component. One survey identified 18 rings in the vicinity of this retroflection [49]. Once formed, Agulhas Rings drift northwestward into the South Atlantic Ocean at a few cm/s speed. Tall seamounts can strongly disrupt the internal balance of these rings and lead to their splitting into two or three parts [3].

3.1.1.2 Mesoscale and Submesoscale Vortices

In the ocean depth, eddies exist which are much smaller than the rings at the surface: their radius is close to the first internal radius of deformation R_{d1} or to the deformation radius corresponding to their vertical structure[1] (i.e., at mid-latitudes, radii smaller than 50 km for mesoscale vortices or even smaller than 10 km for submesoscale vortices, see Table 3.1).

For instance, deep mesoscale anticyclones co-exist with cold-core rings south of the Gulf Stream [137] and smaller, submesoscale, vortices exist in the thermocline of the Sargasso Sea [99].

Among all mesoscale and submesoscale vortices, we focus our attention on eddies of Mediterranean Water, generated on the eastern boundary of the North Atlantic Ocean. The Mediterranean Sea is a concentration basin, forming warm and salty waters (at 300 m depth, in the Alboran Sea, the salinity is above 38 psu). These waters flow out of the Mediterranean basin into the Gulf of Cadiz and adjust there as slope currents (in fact, as three different cores at 600, 800, and 1200 m depths; see Fig. 3.3 (top)). Once these currents have attained their equilibrium depths (near 8°W), they flow quasi-zonally and they encounter the Portimao Canyon, a deep submarine trench. This perturbation on the currents triggers their instability and leads to the formation of intra-thermocline eddies of Mediterranean Water, called meddies (see Fig. 3.3 (bottom)). Meddy radii can vary between 20 and 50 km, their thicknesses between 600 and 1000 m, their azimuthal velocity is close to 0.3–0.5 m/s [5–7, 136], and they present two thermohaline maxima when formed south of the Iberian Peninsula: a temperature maximum ($T \sim 13°C$) near 800 m depth and a salinity maximum ($S \sim 36.6$ psu) near 1200 m depth. Other locations around the Iberian Peninsula are well-known sites of meddy generation: Cape Saint Vincent, the Estramadura Promontory, Cape Ortegal, but due to a change in thermohaline properties of the Mediterranean Water downstream, meddies formed north do not exhibit the double (T,S) structure [119]. Shallower meddies have also been observed [129].

Table 3.1 Characteristics of eddies described in the text

Type of eddy	GS WCR (1)	GS CCR (2)	Agulhas rings (3)	Meddies (3)
Vertical structure	Surface intensified	Whole water column	Surface intensified	Intra-thermocline
Thickness (m)	800	4000	1000	600–1000
Radius (km,*)	60–80	120–150	120–200	20–50
Maximum velocity (m/s)	2.0	1.5	0.8	0.5
Temperature anomaly	7°C	7°C	4°C	4°C
Number formed/year	5–8	5–8	5	15

Notes: (*) based on thermal or haline anomaly, (1) Gulf Stream Warm Core Rings (anticyclonic), (2) Gulf Stream Cold Core Rings (cyclonic), and (3) anticyclonic eddies.

[1] Deformation radii are horizontal length scales over which the effects of planetary rotation and of density stratification on motion are comparable.

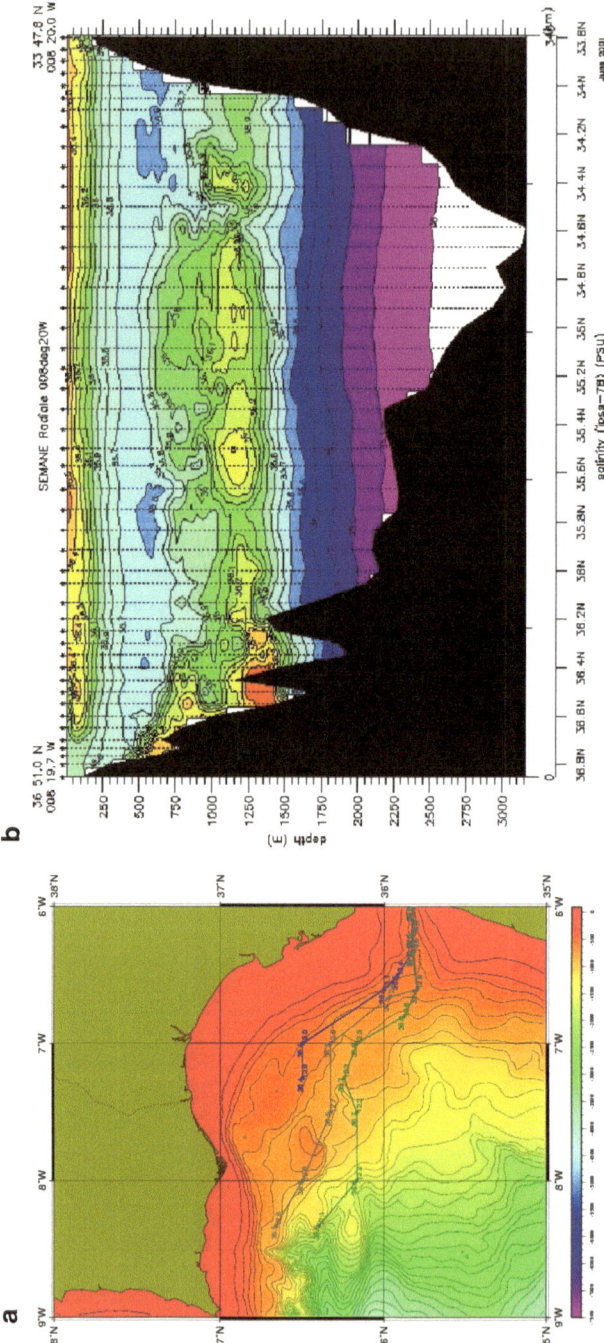

Fig. 3.3 *Top*: the path of the three cores of Mediterranean Water on the Iberian slope, with their respective temperatures and salinities, from Semane95 data; *bottom*: meridional cross-section of salinity along 8°20′W showing the Mediterranean Water cores along the Iberian slope (*left*) and a meddy (lens eddy) detached from the currents (*middle* of the picture), as observed during the Semane2001 experiment

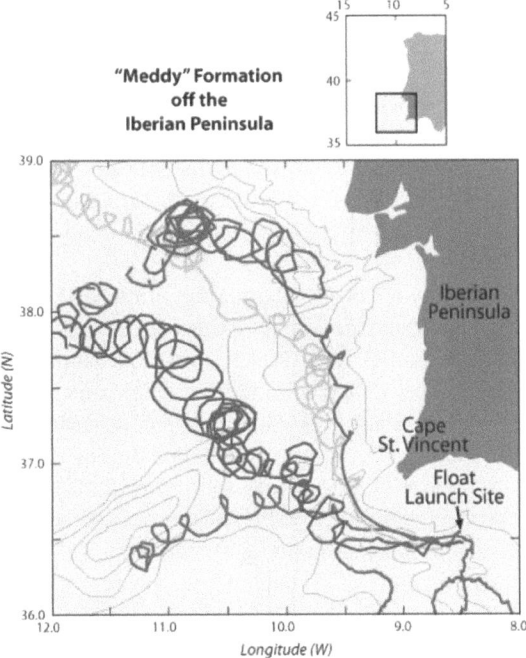

Fig. 3.4 Trajectories of meddies followed by deep float (acoustically tracked) from their site of generation, near Portimao Canyon, from http://www.whoi.edu/ (observing systems: floats and drifters); courtesy Amy Bower, WHOI, Larry Armi, SIO/UCSD and Isabel Ambar, CO/FCUL; see also [19, 20]

Once formed, meddies drift southwestward under the influence of the beta-effect and of the large-scale, baroclinic currents; their typical speed is 3–5 cm/s, though this speed is quite variable (see Fig. 3.4; see [7, 135, 19, 20, 133]). Meddies can undergo destructive encounters with seamount chains (Horseshoe seamounts, mid-Atlantic ridge) which erode them irreversibly [132]. They can also collapse in the deep ocean, when horizontal intrusions of external water masses, acting with double diffusion, erode their core. Nevertheless, an upper bound on meddy lifetime has been set as high as 4 years, and an average lifetime was assessed as 1.7 years [133]. Meddies substantially contribute to the salt flux from the eastern boundary into the North Atlantic Basin: estimates for this eddy salt flux vary between 25 and 100% of the total salt flux in intermediate waters in this region [136, 2, 95].

3.1.2 Generation Mechanisms

3.1.2.1 Barotropic and Baroclinic Instabilities of Parallel Currents

As mentioned above, many oceanic vortices are generated by the instability of intense and narrow parallel flows, e.g., surface-intensified jets in the deep ocean, continental slope currents, and coastal currents. The main processes leading to vortex formation are the barotropic instability of thin jets and the baroclinic instability of currents separating water masses of different densities. Again in most cases, baroclinic instability is more efficient than barotropic instability at producing vortices [39, 38]. Baroclinic instability of surface jets (like the Gulf Stream or the Kuroshio) often has maximum growth rates for wavelengths on the order of $2\pi R_{d1}$. Thus the fastest growing waves would be on the order of 200 km long for the Gulf Stream [56] and about 40 km long for frontal currents on the continental shelf (e.g., the Ushant front; [122]).

Therefore, a large variety of vortices (large rings, mesoscale vortices) can be generated by the same mechanism from different currents. In particular, certain coastal currents lead to the formation of vortex couples (cyclone–anticyclone association): horizontal dipoles in the Sea of Okhotsk [52], baroclinic dipoles from the Mediterranean outflow on the continental slope. The growth of perturbations on intense jets leads to vortices via meander occlusions (corresponding to multiple wave growth and finally to the appearance of long waves, for unforced mean flows). An illustrative animation of meander growth and vortex formation on the Gulf Stream can be found at http://kingfish.coastal.edu/marine/gulfstream.

3.1.2.2 Retroflection, Change of Direction of Currents, Currents Exiting from Straits

Apart from the Agulhas Current, mentioned above, many ocean currents shed vortices as they change direction: the East Madagascar Current south of this island, the North Brazil Current into the NECC (see Fig. 3.5; see also [61, 60]), the Mediterranean outflow at Cape Saint Vincent, or the Navidad (a warm poleward current intensified northwest of Spain in winter) around Cape Ortegal. In the same context, ocean currents exiting from gaps and straits form vortices: the Indonesian through-

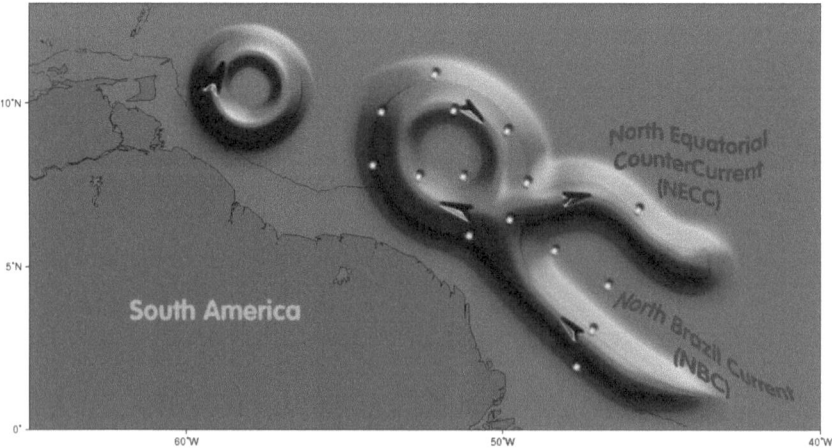

Fig. 3.5 Schematic diagram of ring formation due to the retroflection of the North Brazil Current (from http://www.aoml.noaa.gov/phod/nbc; Sylvia L. Garzoli, personal communication)

flow exiting in the South Indian Ocean forms Teddies (throughflow eddies) and the Loop Current in the Gulf of Mexico forms Loop Current Eddies as it exits from the Straits of Yucatan. The mechanism underlying this phenomenon is momentum conservation: as the current changes direction, momentum is deflected and part of the incoming fluid mass must move in a complementary direction to close the momentum budget [113]. Intrusions of homogeneous fluid in a stratified ocean [62, 4] are also a cause for the formation of vortex lenses.

3.1.2.3 Convection and Geostrophic Adjustment

Several regions of the world ocean (Weddell Sea, Greenland Sea, Labrador Sea, and Gulf of Lion) are sites of deep convection driven by rapid and intense cooling of a weakly stratified water column, preconditioned by wind forcing. Convective chimneys, of size on the order of 50 km, form, containing small-scale vertical plumes (of size 1 km, depending on the heat flux). Via cyclogeostrophic balance,[2] these chimneys correspond to a superposition of a cyclone over an anticyclone which can be baroclinically unstable. When this destabilizes the chimney, meanders grow on the periphery leading to the generation of vortices (of size 5–10 km, often baroclinic dipoles) which propagate away from the chimney spreading the cold water at a speed of 3–5 cm/s [86]. The whole event lasts a few weeks.

3.1.2.4 Topographic Effects on Currents

Seamounts and submarine canyons can trigger the formation of vortices from an intense current, even if this current is stable upstream of the topographic anomaly. In the Bay of Biscay, the Cape Ferret Canyon is recognized as a site of swoddy (slope water oceanic eddy) formation [128]. A deep canyon south of Portugal (Portimao Canyon) is held responsible for triggering the formation of Meddies [35, 144, 36, 143]; see also Fig. 3.3 bottom).

The influence of the New England Seamounts on the Gulf Stream was a debated subject: observations seemed contradictory on the amplification of Gulf Stream meanders downstream of these seamounts. Ezer [51] clarified this question via high-resolution numerical modeling: MKE (mean kinetic energy) increases downstream of the seamounts while EKE (eddy kinetic energy) increases upstream. The propagation speed of meanders is decreased upstream of the seamounts, while topography induces quasi-steady barotropic recirculation cells downstream.

Currents encountering islands can also induce vortices in their lee: for instance, a long-lasting cyclonic vortex, observed in the Alenuihaha Channel between the islands of Hawaii and Maui, was accompanied by intense biological activity. Satellite observations have also revealed that a number of such vortices could form in the lee of the Hawaiian Islands and subsequently merge [54]. One such evolution was observed with satellite altimetry. During this evolution, which lasted 240 days

---

[2] The balance between radial pressure gradient on the one hand and centrifugal and Coriolis force on the other hand.

and counted several consecutive mergers, the vortex radii grew from 50 to 150 km and their turnover period from 3 to 24 days. Eddies have been observed in the lee of other islands, such as the Canary Islands.

3.1.3 Vortex Evolution and Decay

3.1.3.1 Vortex Interactions with Other Currents or with Topography

Vortex–vortex interactions are often nonlinear processes which can lead to the modification of the vortices involved, or even to the disappearance of a vortex. Vortex–vortex interactions include: dipolar pairing, merger, and vertical alignment.

Dipolar pairing is the association of a cyclone and an anticyclone, not necessarily at the same depth. Horizontal and vertical dipoles have been regularly observed in the ocean.

Merger is the process by which two like-signed vortices form a larger vortex, surrounded by filaments or by smaller vortices. Merger has been reported for Gulf Stream rings [74, 50], for Kuroshio rings [142, 168, 167], for Hawaiian eddies [54] and for meddies [145].

Alignment occurs when two like-signed vortices, with cores lying at different depths and at a finite distance initially, tend to form a vertical column in the final state. Vertical alignment has been observed for two East Australian Current eddies [37].

Vortices can also interact with neighboring currents, some of which may be their parent current. Satellite measurements and long-term in situ networks (e.g., the SYNOP experiment) have extensively documented the formation and subsequent merger of Gulf Stream rings with the jet itself [137, 131] and references therein). This occurs when the meander which has created the ring has not completely disappeared after the ring formation or when the newly formed ring has closely approached a following meander of the jet. Rings tend to propagate westward near the Gulf Stream, via a combination of the beta-drift effect and of the interaction with the jet.

Finally, interactions of oceanic vortices with isolated seamounts, with seamount chains, or with island chains have been regularly observed. In particular, Agulhas rings have been observed to encounter the Vema seamount, a tall and narrow seamount in the South Atlantic Ocean [3] and to break under the influence of topographically induced circulation [66, 67].

Deep vortices, such as meddies, have also been observed to interact with seamounts [173, 172, 146, 132, 120, 119]. Meddies can squeeze between tall and narrow seamounts, be eroded, or disintegrate over the Mid-Atlantic Ridge. Meddies can also encounter the continental slope [24] and remain stationary for a significant duration.

3.1.3.2 Vortex Advection, Drift, and Decay

Once formed, vortices rarely remain in their generation region. They can be advected by the mean flow that gave them birth (e.g., the Gulf Stream or the Mediterranean water undercurrents) or they can propagate away from their formation region via self-advection (for instance, in the case of dipolar vortices). In this process, vortices adjust to a stationary form, provided that the surrounding currents are uniform on large enough distances for the adjustment to take place. If self-advection is usually a fast process (dipoles can drift at a speed of 5–8 cm/s), advection by a uniform current is most often slow (drift velocities are often smaller than 1 cm/s).

Another mechanism explaining vortex propagation along a coast is the presence of an opposite-signed image vortex due to the coast. In addition, the continental slope causes vortex squeezing, leading to an asymmetry in vortex structures. These mechanisms are considered responsible for the propagation of North Brazil Current rings along the South American coast [94, 134] and for Loop Current Eddies in the Gulf of Mexico.

To reach the observed drift velocities (3–5 cm/s for meddies, for instance), monopolar (or quasi-monopolar) vortices must be influenced by another factor than uniform currents. Numerical modeling has shown that beta-effects, either planetary or due to baroclinic currents, can induce large drift velocities on vortices. Indeed, ambient vorticity gradients induce an asymmetry in the vortex circulation, which results in their propagation. For instance, the meridional gradient of the Coriolis force induces a meridional drift of vortices, in a direction depending on the vortex polarity [141]. Whatever the vortex polarity, a westward drift, due to the interaction of the vortex and of the planetary beta-effect, is observed [1, 102]. In summary, the planetary beta-effect advects cyclones (resp. anticyclones) northwestward (resp. southwestward) in the northern hemisphere, and conversely in the southern hemisphere: meddies, Agulhas rings clearly confirm this effect. In the ocean, the combination of topographic and baroclinic currents, and of planetary and topographic beta-effects can lead to complex vortex trajectories (see Fig. 3.6).

Atmospheric ventilation is an important process for surface-intensified vortices in regions of strong heat exchange between the atmosphere and ocean (e.g., the Gulf Stream). For instance, warm-core ring 82B was observed to lose 1500 W/m^2 in response to a strong winter storm and ring 82I lost 800 W/m^2 during two consecutive storms [75]. Models show that the response to cooling is governed by Rossby adjustment. Associated with this cooling, convective mixing suppresses the surface signature of the ring [27].

Isopycnal and diapycnal mixing can also affect deep eddies, such as meddies. A long and detailed experiment with Sofar[3] floats and multiple hydrological surveys allowed the investigation of the long-term (2-year long) evolution of the hydrological structure of a meddy [7]. Several mechanisms, lateral intrusions, double diffusion, and turbulence acted to erode the meddy on its top, bottom, and periphery. Lateral intrusions were the most effective to mix heat and salt. The salt content of the meddy core became strongly affected by lateral intrusions reaching the core after 1 year (see Fig. 3.6).

----

[3] SOund Fixing And Ranging.

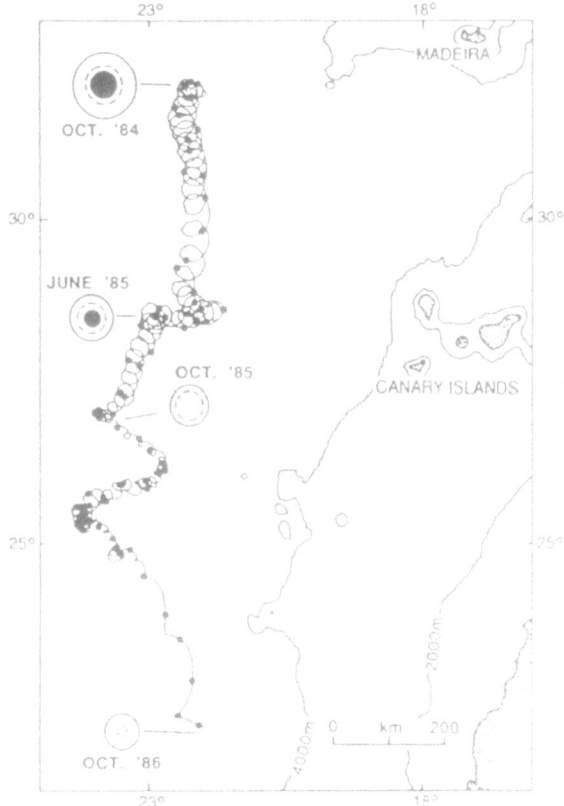

Fig. 3.6 Trajectory of an acoustic float released in a meddy, and remaining in this eddy for 2 years; the structure of the meddy at each survey (core, salinity boundary, and periphery) is indicated on the side; from [7]

3.1.4 Submesoscale Structures and Filaments; Biological Activity

The eddy activity of the ocean is not limited to the coherent vortices described above. Much turbulent activity lies in small-scale, shorter-lived vortices and in filaments (often formed by the interaction between vortices and by the instability of currents (deep-ocean jets, continental slope, and coastal currents). Filaments play an essential role in the vertical transfers to and from the ocean surface. Experiment POMME in the Northeastern Atlantic Ocean had evidenced that the major biological signals (chlorophyll, zooplankton, particles, and pCO_2) were concentrated in frontal and filamentary structures around mesoscale vortices (see Fig. 3.7). Recent numerical simulations of oceanic turbulence at very high resolution have shown that half of the vertical transfers in the upper ocean layers is achieved by elongated filaments, the other half being done by circular filaments and vortices [84].

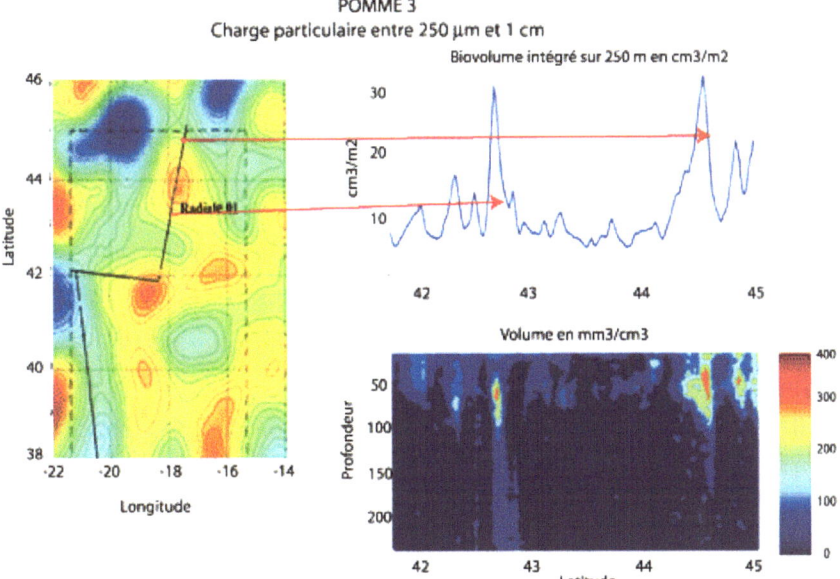

Fig. 3.7 Submesoscale signature during POMME3 experiments: sea level anomaly from a QG forecast model, showing the transect (*left*); vertically integrated content in particles over 250 m (*upper right*) and vertical cross-section (*lower right*). From http://www.lodyc.jussieu.fr/POMME/PROGRAMME/site_pomme_2003

It must nevertheless be stated that biological activity has also been observed in the core of oceanic vortices: cold-core rings of the Gulf Stream initially contain slope water rich in nutrients and plankton; as it decays, this water is replaced by Sargasso Sea water with much lower biological contents [137]. Tropical instability vortices (for a description of these vortices, see [78]), which are synoptic-scale anticyclones, downwell nutrients just below their core, leading to biological production (phytoplankton, zooplankton) just above the nitracline (the level of maximum gradient in nitrate concentration).

3.2 Physical and Mathematical Framework for Oceanic Vortex Dynamics

Many models have been used to describe the dynamics of vortices, depending on the spatial structure or the scale of the effects to be represented: three-dimensional Boussinesq non-hydrostatic models and primitive-equation models have been widely used to study vortex formation, mutual vortex interactions or vortex interactions, with inertia-gravity waves.

Layered models (based on shallow water, frontal geostrophic, and quasi-geostrophic dynamics, hereafter SW, FG, and QG) have been widely used to study vortex stability, interactions, and drift, under the assumption that the ocean (or the atmosphere) can be represented as a stack of homogeneous layers and that vortices are confined in one layer, or in a few of these layers. A central property of these models is conservation of potential vorticity in unforced, non-dissipative flows. Indeed, potential vorticity conjugates many vortex properties (internal vorticity, relation with planetary vorticity, and the vertical stretching of water columns) in a single variable.

3.2.1 Primitive-Equation Model

The primitive equations are the Navier–Stokes equations on a rotating planet, for an incompressible fluid, with Boussinesq and hydrostatic approximations. These dynamical equations are complemented with an equation of state for the fluid and with advection–diffusion equations for temperature and salinity (in the ocean). They are usually written as

$$\frac{Du}{Dt} - fv = \frac{-1}{\rho_0}\partial_x p + F_x + D_x,$$
$$\frac{Dv}{Dt} + fu = \frac{-1}{\rho_0}\partial_y p + F_y + D_y$$

for the two horizontal momentum equations (ρ_0 being an average density),

$$\partial_z p = -\rho g$$

for the hydrostatic balance,

$$\rho = \rho(T, S, p)$$

for the equation of state,

$$\partial_x u + \partial_y v + \partial_z w = 0$$

for the incompressibility equation, and

$$\frac{DT}{Dt} = \kappa_T \nabla^2 T + F_T$$

$$\frac{DS}{Dt} = \kappa_S \nabla^2 S + F_S$$

for the temperature and salinity equations (T is temperature and S is salinity). The Lagrangian advection is three-dimensional $D/Dt = \partial_t + u\partial_x + v\partial_y + w\partial_z$. The

Coriolis parameter is $f = 2\Omega \sin(\theta)$, where Ω is the rotation rate of the Earth and θ is latitude; g is gravity. F_x, F_y and D_x, D_y are the forcing and dissipative terms in the horizontal momentum equations, and F_T, F_S are the source terms in the thermodynamics/tracer equations. The thermal and salt diffusivities are κ_T and κ_S, respectively.

This system is associated with a set of boundary conditions: mechanical, thermal, and haline forcing at the sea surface, interaction with bottom topography, and possible lateral forcing via exchanges between ocean basins.

Primitive equations conserve potential vorticity in adiabatic, inviscid evolutions; this potential vorticity has the form

$$\Pi = (\boldsymbol{\omega} + f\boldsymbol{k}) \cdot \frac{\nabla \rho}{\rho},$$

with $\boldsymbol{\omega} = (-\partial_z v, \partial_z u, \partial_x v - \partial_y u)$.

The primitive equations can be rendered non-dimensional. Non-dimensional numbers quantify the intensity of each physical effect:

- the Rossby number $Ro = U/fL$, where U is a horizontal velocity scale and L a horizontal length scale characterizes the influence of planetary rotation on the motion (this number is the ratio of inertial to Coriolis accelerations),
- the Burger number $Bu = N^2 H^2/f^2 L^2$, where $N^2 = -(g/\rho)\partial_z \rho$ is the Brunt–Väisälä frequency and H is a vertical length scale, indicates the influence of stratification on motion (it is the ratio of buoyancy to Coriolis terms),
- the Reynolds number $Re = UL/\nu$, where ν is viscosity, is the ratio of lateral friction to acceleration and it characterizes the influence of dissipation on motion,
- the Ekman number $Ek = \nu/fH^2$ is the ratio of vertical dissipation to Coriolis acceleration and characterizes the importance of frictional effects at the ocean surface and bottom,
- the aspect ratio of motions, H/L, also indicates how efficiently planetary rotation and ambient stratification have confined motions in the horizontal plane.

These non-dimensional numbers are used to derive the simplified equation systems (shallow-water and quasi-geostrophic models). In particular, for unforced, non-dissipative motions, a small Rossby number (associated with small aspect ratio of the motion) indicates that the Coriolis acceleration balances the horizontal pressure gradient:

$$fv = \frac{1}{\rho_0}\partial_x p$$
$$fu = \frac{-1}{\rho_0}\partial_y p.$$

These equations are called the geostrophic balance. Using now the hydrostatic balance, and under the same conditions, we obtain the thermal wind relations

$$f \partial_z u = \frac{-g}{\rho_0} \partial_x \rho,$$

$$f \partial_z v = \frac{g}{\rho_0} \partial_y \rho,$$

which indicates that the vertical shear of horizontal velocity is then related to the horizontal density gradients.

The primitive-equation model has been used for the study of vortex generation by deep ocean jets or by coastal currents.

Along the continental shelf from the Florida Straits to Cape Hatteras, the Gulf Stream is a frontal current and it can undergo frontal baroclinic instability, leading to the formation of meanders and cyclones. With a primitive-equation model, Oey [115] showed that the relative thickness of the upper ocean layer and the distance of the front from the continental slope govern the frontal baroclinic instability. Chao and Kao [26] evidenced successive barotropic and baroclinic instabilities on this current and the formation of anticyclones. To analyze the formation of meanders and rings in the Gulf Stream region east of Cape Hatteras, Spall and Robinson [147] used a primitive-equation, open-ocean model, and they showed that bottom topography plays an important role in the structure of the deep flow. Warm-core ring formation results from differential horizontal advection of a developed meander, while cold-core ring formation involves geostrophic and ageostrophic horizontal advection, vertical advection, and baroclinic conversion.

With a primitive-equation model, Lutjeharms et al. [93] studied the formation of shear edge eddies from the Agulhas Current along the Agulhas Bank. These eddies, with a diameter of 50–100 km, are prevalent in the Agulhas Bank shelf bight as observed, and their leakage may trigger the detachment of cyclones from the tip of the Agulhas Bank. These cyclones have sometimes been observed to accompany the detachment of Agulhas rings from the Agulhas Current.

More recently, the primitive-equation model was used for the study of ocean surface turbulence, vertical motions and the coupling of physics with biology, via submesoscale motions. Levy et al. [88] modeled jet instability at very high resolution and showed that submesoscale physics reinforce the mesoscale eddy field. Submesoscale structures (filaments) are associated with strong density and vorticity gradients and are located between the eddies. They also induce large vertical velocities, which inject nutrients in the upper ocean layer. This study was complemented by that of Lapeyre and Klein [84] who showed that elongated filaments are more efficient than curved filaments at injecting nutrients vertically.

3.2.2 The Shallow-Water Model

3.2.2.1 Equations and Potential Vorticity Conservation

At eddy scale or even at the synoptic scale (a few hundred kilometers horizontally), the ocean can be modeled as a stack of homogeneous layers in which the motion is essentially horizontal (due to Coriolis force and stratification). In each layer, horizontal homogeneity leads to vertically uniform horizontal velocities. The shallow-water equations are obtained by integrating the horizontal momentum and the incompressibility equations over each layer thickness. Here, we write the shallow-water equations in polar coordinates for application to vortex dynamics (u_j is radial velocity and v_j is azimuthal velocity):

$$\frac{Du_j}{Dt} - f v_j = \frac{-1}{\rho_j}\partial_r p_j + F_{rj} + D_{rj}$$

$$\frac{Dv_j}{Dt} + f u_j = \frac{-1}{r\rho_j}\partial_\theta p_j + F_{\theta j} + D_{\theta j}$$

$$\frac{Dh_j}{Dt} + h_j \nabla \cdot \mathbf{u_j} = \frac{Dh_j}{Dt} + \frac{h_j}{r}(\partial_r(r u_j) + \partial_\theta v_j) = 0, \qquad (3.1)$$

with

$$\frac{D}{Dt} = \partial_t + u_j\partial_r + (v_j/r)\partial_\theta.$$

Here p_j, h_j, ρ_j, F_j, and D_j are pressure, local thickness, density, body force, and viscous dissipation, respectively in layer j (j varying from 1 at the surface to N at the bottom); $f = f_0 + \beta y$ is the expansion of the spherical expression of f on the tangential plane to Earth at latitude θ_0. The local and instantaneous thickness is $h_j = H_j + \eta_{j-1/2} - \eta_{j+1/2}$, where H_j is the thickness of the layer at rest and $\eta_{j+1/2}$ is the interface elevation between layer j and layer $j+1$ due to motion. We choose to impose a rigid lid on the ocean surface ($\eta_{1/2} = 0$) and the bottom topography is represented by $\eta_{N+1/2} = h_B(x, y)$ (see Fig. 3.8). Finally, the hydrostatic balance is written as $p_j = p_{j-1} + g(\rho_j - \rho_{j-1})\eta_{j-1/2}$.

An essential property of these equations is layerwise potential vorticity conservation in the absence of forcing and of dissipation ($F_j = D_j = 0$). By taking the curl of the momentum equations, and by substituting the horizontal velocity divergence in the continuity equation, Lagrangian conservation of layerwise potential vorticity Π_j is obtained:

$$\frac{d\Pi_j}{dt} = 0, \qquad \Pi_j = \frac{\zeta_j + f_0 + \beta y}{h_j}, \qquad (3.2)$$

with $\zeta_j = (1/r)[\partial_r(r v_j) - \partial_\theta u_j]$ the relative vorticity.
For vortex motion, it is more convenient to introduce the PV anomaly with respect to the surrounding ocean at rest. For instance, in the case of f-plane dynamics

$$Q_j = \Pi_j - \Pi_j^0 = \frac{\zeta_j + f_0}{h_j} - \frac{f_0}{H_j} = \frac{1}{h_j}\left(\zeta_j - f_0\frac{\delta\eta_j}{H_j}\right),$$

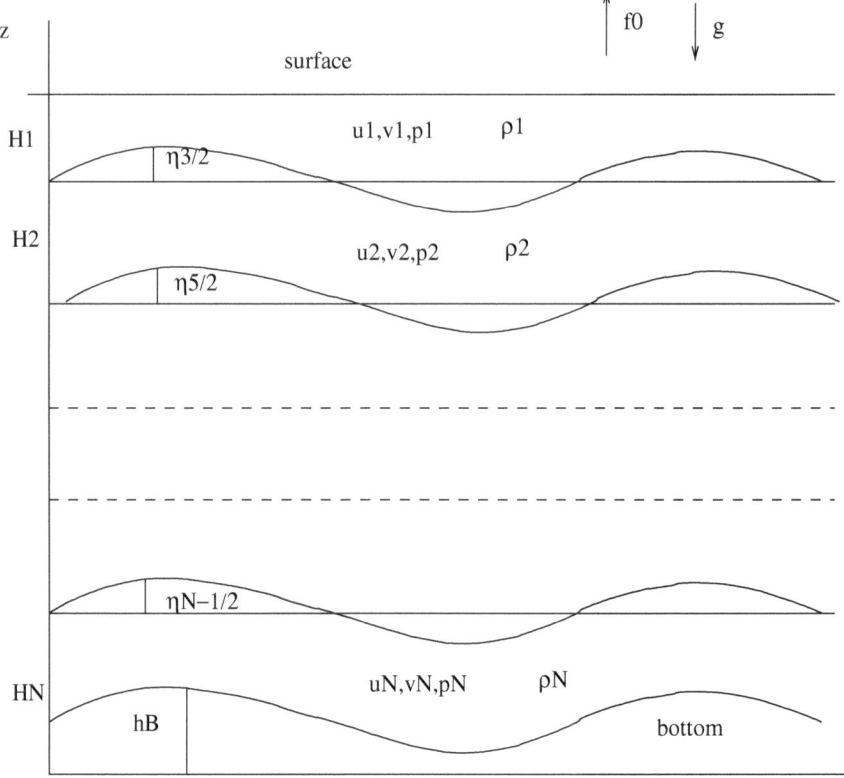

Fig. 3.8 Sketch of a N-layer ocean for the shallow-water model

where $\delta\eta_j = h_j - H_j$ is the vertical deviation of isopycnals across the vortex. Obviously, the PV anomaly is then conserved. On the beta-plane, one usually does not include planetary vorticity in the PV anomaly, which is then not conserved [108].

To evaluate the potential vorticity contents of each layer, we restore the forcing and dissipation terms, so that

$$\frac{d\Pi_j}{dt} = \frac{1}{h_j} \left[\frac{1}{r} \partial_r (r (F_{\theta j} + D_{\theta j})) - \frac{1}{r} \partial_\theta (F_{rj} + D_{rj}) \right].$$

Now

$$\frac{d\Pi_j}{dt} = \partial_t \Pi_j + \boldsymbol{u}_j \cdot \boldsymbol{\nabla}\Pi_j = \partial_t \Pi_j + \boldsymbol{\nabla} \cdot [\boldsymbol{u}_j \Pi_j]$$

using the non-divergence of horizontal velocity. Therefore, if we integrate the relation above on the volume of layer j, we have

$$\frac{d}{dt} \int\int_{S_j} \Pi_j h_j dS = \oint_{C_j} (\boldsymbol{F}_j + \boldsymbol{D}_j) \cdot \boldsymbol{dl}_j,$$

where C_j is the boundary of S_j (see [64, 65, 109]). Thus, the potential vorticity contents in layer j vary when forcing or dissipation is applied at the boundary of the layer. The equation for the potential vorticity anomaly is the following:

$$\frac{d}{dt} \int\int_{S_j} Q_j h_j dS = -f \frac{dV_j}{dt} + \oint_{C_j} (\boldsymbol{F}_j + \boldsymbol{D}_j) \cdot \boldsymbol{dl}_j,$$

where V_j is the volume of layer j [109]. Thus, the potential vorticity anomaly contents can change when this volume varies (e.g., via diapycnal mixing) or when forcing or dissipation occurs at the boundary of the layer. This "impermeability theorem" has important consequences for flow stability (see also [110]).

For isopycnic layers which intersect the surface, Bretherton [21] has shown that "a flow with potential [density] variations over a horizontal and rigid plane boundary may be considered equivalent to a flow without such variations, but with a concentration of potential vorticity very close to the boundary." In particular, Boss et al. [16] show that an outcropping front corresponds to a region of very high potential vorticity, conditioning the instabilities which can develop on this front.

3.2.2.2 Velocity–Pressure Relations and Inversion of Potential Vorticity

The prescription of the potential vorticity distribution characterizes the eddy structure, but one needs to know the associated velocity field to determine how the eddy will evolve. To do so, one needs a diagnostic relation between pressure (or layer thickness) and horizontal velocity, to invert potential vorticity into velocity. In the shallow-water model, such a relation does not always exist. One important instance where it does is the case of circular eddies.

It can be easily shown that axisymmetric and steady motion in a circular eddy obeys a balance between radial pressure gradients, Coriolis and centrifugal accelerations, called cyclogeostrophic balance; this is obtained by simplifying the shallow-water equations above with $\partial_t = 0$, $\partial_\theta = 0$, $v_r = 0$ (see [40])

$$-\frac{v_\theta^2}{r} - f_0 v_\theta = \frac{-1}{\rho} \frac{dp}{dr}. \tag{3.3}$$

In this case, inversion of potential vorticity into velocity leads to a nonlinear ordinary differential equation which can be solved iteratively, if the centrifugal term is weak compared to the Coriolis term.

This equation can be put in non-dimensional form with the Rossby number $Ro = U/f_0 R$ and the Burger number $Bu = g'H/f_0^2 R^2$ with $U, R, \Delta H, H$ scaling the eddy azimuthal velocity, radius, and thickness and the upper layer thickness:

$$\text{Ro} \frac{v_\theta^2}{r} + v_\theta = \frac{Bu}{Ro} \frac{\Delta H}{H} \frac{d\eta}{dr}. \tag{3.4}$$

Note that this balance introduces an asymmetry between cyclones and anticyclones (see also [23]).

For small Rossby numbers, geostrophic balance holds:

$$U = \frac{g'\Delta H}{f_0 R} \quad \text{and} \quad \frac{\Delta H}{H} = \frac{Ro}{Bu},$$

while for Rossby numbers of order unity or larger, horizontal velocity scales on pressure gradient via the centrifugal term (cyclogeostrophic balance) and

$$U = \sqrt{g'\Delta H} \quad \text{and} \quad \frac{\Delta H}{H} = \frac{Ro^2}{Bu}.$$

Lens eddies are defined by large vertical deviations of isopycnals $\Delta H / H \sim 1$ or $Ro \sim Bu$, and they are described by the full shallow-water equations (or by frontal geostrophic equations, see below). Quasi-geostrophic vortices correspond to smaller deviations of isopycnals, i.e., $\Delta H / H << 1$ or $Ro << 1, Bu \sim 1$.

In fact, the cyclogeostrophic balance is the f-plane, axisymmetric version of the gradient wind balance. To obtain the gradient wind balance, one starts from the horizontal velocity divergence equation. Calling $\Delta_j = \frac{1}{r}\partial_r r u_j + \frac{1}{r}\partial_\theta v_j$ the horizontal divergence, this equation is

$$\frac{d\Delta_j}{dt} + \Delta_j^2 - 2J(u_j, v_j) - f\zeta_j + \beta \cos(\theta) u_j = -\frac{1}{\rho_j}\nabla^2 p_j + \nabla \cdot [\mathbf{F}_j + \mathbf{D}_j],$$

where $J(a, b) = \frac{1}{r}[\partial_r a \partial_\theta b - \partial_r b \partial_\theta a]$ is the Jacobian operator. In the absence of forcing and dissipation, if the Rossby number is small, the advection of horizontal velocity divergence and the squared divergence are smaller than the other terms. The equation becomes then

$$2J(u_j, v_j) + f\zeta_j - \beta \cos(\theta) u_j = \frac{1}{\rho_j}\nabla^2 p_j,$$

which is the gradient wind balance. On the f-plane, this equation is

$$2J(u_j, v_j) + f_0\zeta_j = \frac{1}{\rho_j}\nabla^2 p_j,$$

which, for a circular eddy, is the divergence of the cyclogeostrophic balance.

For eddies which are not circular, the gradient wind balance provides a diagnostic relation between velocity and pressure, which must be solved iteratively. Writing this balance

$$\zeta_j = \frac{1}{f_0\rho_j}\nabla^2 p_j - \frac{2}{f_0}J(u_j, v_j)$$

the first term on the right-hand side of the equation is called the geostrophic relative vorticity, and the second term is a first-order approximation (in Rossby number) of the ageostrophic relative vorticity. At first order in the iterative solution procedure, this balance is written as

$$\zeta_j = \frac{1}{f_0 \rho_j} \nabla^2 p_j - \frac{2}{f_0^2 \rho_j} J(\partial_x p_j, \partial_y p_j),$$

using in the Jacobian operator geostrophic balance to replace velocity into pressure gradient. This relation is a Monge–Ampère equation which has a limited solvability. If a solution exists, the potential vorticity distribution can be inverted into pressure and then into velocity.

On the f-plane and in a one-and-a-half layer reduced gravity model, for a circular, anticyclonic, lens eddy, with zero potential vorticity and radius R, potential vorticity can be easily inverted into pressure (height) and velocity fields. In this case, relative vorticity is equal to $-f_0$ and azimuthal velocity is equal to $-f_0 r/2$. The cyclogeostrophic balance leads to

$$h(r) = \frac{f_0^2}{8g'}(R^2 - r^2),$$

where R is the eddy radius. The central thickness is $h(0) = f_0^2 R^2/(8g')$.

Another instance where potential vorticity is easily inverted is the case of a circular eddy with constant potential vorticity $q > 0$ inside radius R and constant potential vorticity q' outside. Assuming here geostrophic balance, the layer thickness satisfies the equation

$$\frac{d^2h}{dr^2} + \frac{1}{r}\frac{dh}{dr} - \frac{f_0 q}{g'}h + \frac{f_0^2}{g'} = 0$$

for $r \leq R$. The inner solution is $h(r) = (f_0/q) + h_0 I_0(r\sqrt{f_0 q/g'})$, where I_0 is the modified Bessel function of the first kind of order zero. The equation for the layer thickness outside is similar to that inside the eddy, and the outer solution is $h(r) = (f_0/q') + h_1 K_0(r\sqrt{f_0 q'/g'})$, where K_0 is the modified Bessel function of the second kind of order zero. The two constants h_0 and h_1 are obtained by matching h and the azimuthal velocity $(g'/f_0)dh/dr$ at $r = R$:

$$\frac{f_0}{q} + h_0 I_0\left(R\sqrt{\frac{f_0 q}{g'}}\right) = \frac{f_0}{q'} + h_0 I_0\left(R\sqrt{\frac{f_0 q'}{g'}}\right)$$

$$h_0\sqrt{q} I_1\left(R\sqrt{\frac{f_0 q}{g'}}\right) = -h_1\sqrt{q'} K_1\left(R\sqrt{\frac{f_0 q'}{g'}}\right),$$

where I_1 and K_1 are modified Bessel function of the first and second kinds of order one. Obviously, such calculations must be performed numerically when centrifugal terms are inserted in the velocity–pressure relation.

3.2.2.3 Flow Stationarity

The cyclogeostrophic solution presented above shows that a circular vortex remains stationary on the f-plane. But this case is not the only stationary solution of the shallow-water equations. For instance, on the f-plane, a steadily rotating vortex with constant rotation rate Ω, obeys the following equations (in the absence of forcing and of dissipation)

$$u'_j \, \partial_r u'_j + \left(v'_j/r\right) \partial_\theta u'_j - f v'_j = \frac{-1}{\rho_j} \partial_r p'_j$$

$$u'_j \, \partial_r v'_j + \left(v'_j/r\right) \partial_\theta v'_j + f u'_j = \frac{-1}{r\rho_j} \partial_\theta p'_j$$

$$\partial_r \left(r h_j u'_j\right) + \partial_\theta \left(r h_j v'_j\right) = 0,$$

where $u'_j = u_j$, $v'_j = v_j - \Omega r$, $h'_j = h_j$, $p'_j = p_j + \frac{\Omega^2 r^2}{2}$ and $f = f_0 + 2\Omega$. Note that these equations can also be written as

$$\left(\zeta'_j + f\right) \mathbf{k} \times \mathbf{u}'_j + \nabla \left[\frac{p'_j}{\rho_j} + \frac{1}{2}\left(\left(u'_j\right)^2 + \left(v'_j\right)^2\right) \right] = 0$$

and

$$\nabla \cdot [h_j \, \mathbf{u}'_j] = 0.$$

Setting $B'_j = \left(p'_j/\rho_j\right) + \left(\left(u'_j\right)^2 + \left(v'_j\right)^2\right)/2$ and eliminating velocity between both equations, the condition for steadily rotating shallow-water flows is

$$J\left(B'_j, \Pi'_j\right) = 0,$$

with $\Pi'_j = (\zeta''_j + f)/h_j$. This leads to $B'_j = F\left(\Pi'_j\right)$.
Note also that the non-divergence of mass transport implies the existence of a transport streamfunction ψ_j such that $h_j u'_j = -(1/r)\partial_\theta \psi_j$, $h_j v'_j = \partial_r \psi_j$. The momentum equations are then

$$\Pi'_j \nabla \psi_j = -\nabla B'_j = -\nabla \Pi'_j F'\left(\Pi'_j\right),$$

and therefore

$$\nabla \psi_j = -\nabla \Pi'_j F'\left(\Pi'_j\right)/\Pi'_j = \nabla \left(G\left(\Pi'_j\right)\right),$$

thus relating transport streamfunction and potential vorticity.

An example of steadily rotating shallow-water vortex is the rodon, a semi-ellipsoidal surface vortex on the f-plane in a one-and-a-half layer model. This vortex was used to model Gulf Stream rings.

On the beta-plane, vortex stationarity is conditioned by the "no net angular momentum theorem," originally presented in Flierl et al. [59] and later developed by Flierl [55]. If the vortex is vertically confined between two isopycnals, it will remain stationary on the beta-plane (in the absence of forcing and of dissipation) if its net angular momentum vanishes to avoid a meridional imbalance in Rossby force (Coriolis force acting on the azimuthal motion). This condition is expressed mathematically as:

$$\beta \int \int \Psi \, dxdy = 0,$$

where Ψ is the transport streamfunction associated to the vortex.

Note that this condition can also be obtained by canceling the drift speed for lens eddies on the beta-plane calculated by Nof [111, 112] and Killworth [79]

$$c = -\frac{\beta}{f} \frac{\int \int \Psi dxdy}{\int \int h dxdy}.$$

3.2.2.4 Rayleigh-Type Stability Conditions for Vortices in the Shallow-Water Model

The former two paragraphs have described the structure of isolated, stationary vortices in the shallow-water model. They have not dealt with conditions for their stability. Ripa [138, 139] derived stability conditions for circular vortices (on the f-plane) and for parallel flows, with a variational method. Stable solutions were characterized as minima of pseudo-energy (energy added to functionals of potential vorticity and to angular momentum).

Due to potential vorticity conservation in the absence of forcing and of dissipation, functionals of potential vorticity are invariants of the flow:

$$I[F] = \sum_{j=1}^{N} \int \int h_j F_j(\Pi_j) \, rdrd\theta,$$

with $\Pi_j = (f + V_j/r + dV_j/dr)/H_j$.

Total energy is also conserved under the same conditions:

$$E = \frac{1}{2} \int \int \left[\sum_{j=1}^{N} h_j \left(u_j^2 + v_j^2 \right) + \sum_{j=1}^{N'} g'_j \eta_{j+1/2}^2 \right] rdrd\theta,$$

with $N' = N$ for reduced gravity flows and $N' = N - 1$ for flat bottom oceans. Angular momentum is conserved for unforced, inviscid flows

$$A = \int \int \sum_{j=1}^{N} h_j \left(r v_j + \frac{1}{2} f r^2 \right) r dr d\theta.$$

Starting from an axisymmetric flow in cyclogeostrophic balance

$$U_j = 0, \ V_j = V_j(r), \ H_j = H_j(r), \ P_j = P_j(r),$$

if all small perturbations $[u', v', h']$ satisfy

$$\delta S = S[U + u', H + h'] - S[U, H] > 0,$$

with $S = E - \sigma A - I[F]$ (σ a constant), then the flow is stable. The first variation $\delta^{(1)}S$ will vanish if $F_j - \Pi_j d F_j / d\Pi_j = \frac{1}{2} V_j^2 - \sigma \left(V_j r - \frac{1}{2} f r^2 \right) + P_j$ in each layer. Then, the second variation of S will be

$$\delta^{(2)}S = \frac{1}{2} \int \int \left[\sum_{j=1}^{N} H_j \left((u')_j^2 + (v')_j^2 \right) + (V_j - \sigma r) \{ 2(v')_j (h')_j + \frac{\xi_j^2}{d\Pi_j/dr} \} \right] + \sum_{j=1}^{N'} g'_j (\eta')_{j+1/2}^2] \, r dr d\theta.$$

Some algebra (see [138]) is needed to convert $\delta^{(2)}S$ into a simpler form, which is positive definite (implying a stable flow) if the following conditions are satisfied:

1) if there exists $\sigma \neq 0$ such that

$$\frac{V_j - \sigma r}{d\Pi_j/dr} < 0$$

for all r and for all $j = 1, \ldots, N$, and
2) if $G_{ij}(\sigma)$ is positive definite with

$$G_{ii} = g'_i - \lambda_i - \lambda_{i+1}, \ G_{i-1,i} = \lambda_i, \ G_{i+1,i} = \lambda_{i+1},$$

and $G_{ij} = 0$ otherwise, with $\lambda_j = (V_j - \sigma r)^2 / H_j$, then the flow is stable.

The first condition is derived from the Rayleigh inflection point theorem [130], the second condition is a subcriticality condition.

Three examples of applications are

- the two-dimensional flow where there is no subcriticality condition, and where the first condition is equivalent to the Rayleigh stability condition by choosing σ out of the range of values of $V(r)/r$.
- the one-and-a-half layer reduced gravity flow, for which the subcriticality condition is $(V - \sigma r)^2 < g'H$.
- the two-layer (flat bottom) flow, for which this condition becomes

$$\frac{(V_1 - \sigma r)^2}{g'H_1} + \frac{(V_2 - \sigma r)^2}{g'H_2} < 1.$$

3.2.2.5 Balanced Dynamics

The shallow-water model allows both fast and slow motions (e.g., inertia-gravity waves versus vortical motions). For slow motions, relative acceleration is small compared to Coriolis accelerations, and the divergent flow remains weak at all times. In the shallow-water model, a usual decomposition of the velocity in streamfunction ψ and velocity potential χ is

$$u = k \times \nabla \psi + \nabla \chi.$$

In the one-and-a-half layer reduced-gravity model, relative vorticity is $\zeta = \nabla^2 \psi$ and the horizontal velocity divergence is $D = \nabla^2 \chi$. Their evolution equations are written as

$$\partial_t \zeta + fD = -\nabla \cdot (v\zeta)$$
$$\partial_t D + g\nabla^2 h - f\zeta = 2J(u, v) - \nabla \cdot (vD).$$

Slow motions are characterized by mostly rotational flows, i.e., $\chi \sim O(Ro)\psi$. When this condition is inserted in the divergence equation, the remaining terms at $O(Ro)$ form the Bolin–Charney balance [15, 30]. On the f-plane, this balance is written as

$$f_0\nabla^2\psi + 2J(\partial_x\psi, \partial_y\psi) = g\nabla^2 h,$$

which is the gradient wind balance presented above (further details are available in [100]).

The problem of separating these two types of motions in numerical weather predictions, and in particular of suppressing transient, fast motion (often gravity waves generated by unbalanced initial conditions), has been the subject of many studies since the 1950s (e.g., [30, 15, 124, 68, 87, 89, 69, 162]). Many balanced equation models have been developed and applied to vortex dynamics and to oceanic turbulence (e.g., [103, 106, 169–171, 105]). Recently, original systems of balanced equations or balance conditions were derived for the shallow-water model: first, the slaving principle of Warn et al. [165] and then the hierarchy of balance conditions of

Mohebalohojeh and Dritschel which relate to the work of McIntyre and Norton [97].
Both systems of equations are convenient for vortex dynamics (see also a recent
review in [98]).

Mesoscale oceanic motions such as long-lived eddies mostly obey the Bolin–
Charney balance, and thus they have been studied in various kinds of geostrophic
models: balanced equations, frontal geostrophic, generalized geostrophic, or quasi-
geostrophic models, two of which are now presented.

3.2.3 Frontal Geostrophic Dynamics

When $Ro \ll 1$, the shallow-water equations have been expanded in this small
parameter to express horizontal velocity in terms of height in a variety of manners.
In particular, when $Ro \sim Bu$, lens eddies which are not too intense are described
by a set of equations called the frontal geostrophic equations. These equations have
been derived mostly in the context of one-and-a-half layer reduced gravity flows
[41, 42, 148, 149] and of two-layer flows [43, 157, 155, 11–14, 77].

In the one-and-a-half layer reduced gravity model, frontal-geostrophic equations
describe the time evolution of the layer thickness h (since horizontal variations of
this thickness occur on synoptic scales, vortex stretching dominates relative vorticity
in potential vorticity):

$$\partial_t h + J\left(h\nabla^2 h + \frac{1}{2}|\nabla h|^2\right) = 0.$$

In the two-layer model, when the flow is surface-intensified, a thin surface layer is
the usual assumption. Then the lower layer is quasi-geostrophic:

$$\partial_t h + J\left(p + h\nabla^2 h + \frac{1}{2}|\nabla h|^2\right) = 0$$

$$\partial_t[\nabla^2 p + h] + J(p, \nabla^2 p + h + h_b) + \beta \partial_x p = 0,$$

where h is the upper-layer thickness, p is the lower-layer pressure, and h_b is bottom
topography elevation.

Note that, for bottom-intensified flows over topography, Swaters [154, 156] has
derived the dynamical equations which are only quadratic in the variables

$$\nabla^2 \eta_t + J(h + \eta, h_b) + J(\eta, \nabla^2 \eta) = 0$$

$$h_t + J(\eta + h_b, h) = 0,$$

where η is the sea surface elevation, h is the bottom layer thickness, and h_b the
bottom topography elevation, as above.

Frontal geostrophic models have often been used to study the formation of vortices from unstable surface or bottom flow, and vortices in turbulent flows. The
surface frontal geostrophic equations imply a different behavior of cyclones and of

anticyclones. Indeed, it was shown that anticyclones are more stable than cyclones on the f-plane and on the beta-plane propagate westward faster than cyclones.

3.2.4 Quasi-geostrophic Vortices

3.2.4.1 Model Equations

The quasi-geostrophic model is derived from the primitive equations (in continuous stratification) or from the shallow-water equations (in layerwise form) assuming small Rossby number (weak relative acceleration compared to Coriolis accelerations), order unity Burger number (small vertical deviations of isopycnals), and small height of bathymetry, compared to the bottom layer thickness. It is also assumed that the latitudinal variation of the Coriolis parameter remains moderate (planetary scales are excluded). The original derivation of the quasi-geostrophic model is due to Charney [28, 29].

Since relative acceleration and beta-effect are weak, the flow is nearly in geostrophic equilibrium (hence the name "quasi-geostrophic"); therefore, at zeroth order in Rossby number $Ro = U/f_0 L$ (L being a horizontal length scale), the flow is horizontally non-divergent:

$$u = u^{(0)} + Ro u^{(1)} + \cdots, \quad v = v^{(0)} + Ro v^{(1)} + \cdots$$

$$u^{(0)} = -\frac{1}{\rho f_0} \partial_y p, \quad v^{(0)} = \frac{1}{\rho f_0} \partial_x p, \quad \partial_x u^{(0)} + \partial_y v^{(0)} = 0,$$

thus defining a streamfunction $\psi = p/(\rho f_0)$.

The vertical velocity gradient will equilibrate the horizontal flow divergence at first order in Rossby number

$$w^{(0)} = 0, \ \partial_z w^{(1)} = -[\partial_x u^{(1)} + \partial_y v^{(1)}].$$

Here, as in the shallow-water model, momentum and vorticity advection are performed by the horizontal flow only.[4] Therefore, calculating the relative vorticity equation and substituting horizontal velocity divergence as in the shallow-water equations, one also obtains potential vorticity conservation in the absence of forcing and of dissipation. In layerwise form, this equation is

$$\frac{dq_j}{dt} = 0 = \partial_t q_j + u_j^{(0)} \partial_x q_j + v_j^{(0)} \partial_y q_j = \partial_t q_j + J(\psi_j, q_j).$$

Note that the quasi-geostrophic potential vorticity is the first-order term in a Rossby number expansion of the shallow-water potential vorticity *anomaly*.

---

[4] In the continuously stratified quasi-geostrophic model, this also holds, contrary to the PE model.

To determine the expression of quasi-geostrophic potential vorticity, we start from a non-dimensional $\delta \bar{\Pi}_j$:

$$\delta \bar{\Pi}_j = \frac{H_j}{f_0} \Pi_j - 1 = \frac{1}{f_0 h_j} [H_j(\zeta_j + f) - f_0 h_j].$$

Recalling that

$$h_j = H_j \left(1 + \frac{Ro}{Bu} \delta \bar{\eta}_j\right),$$

and

$$f = f_0(1 + R_\beta \bar{y}),$$

with

$$R_\beta = \beta L / f_0 \leq Ro, \quad \bar{y} = y/L, \quad \delta \bar{\eta}_j = \delta \eta_j / H_j,$$

and setting $\zeta_j / f_0 = Ro \bar{\zeta}_j$, one obtains

$$\delta \bar{\Pi}_j \sim Ro \left[\bar{\zeta}_j + \frac{R_\beta}{Ro} \bar{y} - \frac{1}{Bu} \delta \bar{\eta}_j\right] + O(Ro^2)$$

so that, naturally, in non-dimensional form $\bar{q}_j = (1/Ro)\delta \bar{\Pi}_j$.
Finally, calling $\bar{\beta} = R_\beta / Ro$, $\bar{\psi}_j = \psi_j / UL$, and expressing relative vorticity and the stretching of water columns (vortex stretching) in terms of streamfunction, via

$$\bar{\zeta}_j = \nabla^2 \bar{\psi}_j$$

and

$$\delta \bar{\eta}_j / Bu = F_{j,j-1/2}[\bar{\psi}_j - \bar{\psi}_{j-1}] + F_{j,j+1/2}[\bar{\psi}_j - \bar{\psi}_{j+1}],$$

the non-dimensional quasi-geostrophic potential vorticity is written as

$$\bar{q}_j = \nabla^2 \bar{\psi}_j - F_{j,j-1/2}[\bar{\psi}_j - \bar{\psi}_{j-1}] + F_{j,j+1/2}[\bar{\psi}_j - \bar{\psi}_{j+1}] + 1 + \bar{\beta}y,$$

with $F_{j,j+1/2} = f_0^2 L^2 / g'_{j+1/2} H_j$ (here 1 stands for f_0).
A rigid lid on the upper layer cancels $F_{1,1/2}$ while bottom topography is taken into account by replacing $F_{N,N+1/2}[\bar{\psi}_N - \bar{\psi}_{N+1}]$ by $-h_b/H_N$ (dimensionally by $-f_0 h_b/H_N$).
When $f = f_0$, the dynamics are those of the f-plane; when $f = f_0 + \beta y$, beta-plane dynamics are studied.

3.2.4.2 Equations for Continuous Stratification

Note that potential vorticity conservation can also be expressed in terms of stream-function in the continuously stratified quasi-geostrophic model as

$$[\partial_t + J(\bar{\psi}, \cdot)]\,\bar{q} = 0,$$

where (again in non-dimensional form)

$$\bar{q} = \nabla^2 \bar{\psi} + \partial_z \left(\frac{f_0^2 L^2}{N^2 H^2} \partial_z \bar{\psi} \right) = 0,$$

and N^2 is the squared Brunt–Väisälä frequency.

Usually the stratification operator $\partial_z \left(\frac{f_0^2 L^2}{N^2 H^2} \partial_z \right)$ is diagonalized to provide vertical eigenmodes (see more details in [123] or in [23]). The modal and layerwise descriptions of motions are formally equivalent.

In fact, the conservation and impermeability theorems for potential vorticity, were first derived in continuously stratified quasi-geostrophic flows, [64, 44, 65]. The (more recent) shallow-water version of these theorems was presented in Sect. 3.2.2.

In the quasi-geostrophic framework, these theorems state that even in the presence of diabatic heating and frictional or other forces, there can be no net transport of potential vorticity across any isentropic surface in the atmosphere (or across any isopycnic surface in the ocean), and that potential vorticity can neither be created nor destroyed within a layer bounded by two isentropic (isopycnic) surfaces. Consequently, it can be created or destroyed at places (if any) where the layer ends laterally. This concerns isopycnic layers which ventilate, for instance, or which intersect the sea floor.

Another essential principle concerning potential vorticity is its invertibility, i.e., the possibility to recover the flow structure from the potential vorticity distribution, as long as limits of centrifugal, static instabilities or a change in sign of the quantity (absolute vorticity + strain rate) are not reached [101]. This invertibility has been studied at length by McWilliams and Gent [104], Hoskins et al. [69], McIntyre and Norton [96] (see also above, "shallow-water model" and "balanced models").

In the quasi-geostrophic model, invertibility of potential vorticity into stream-function is possible for all physically realistic problems (which are thus mathematically well-posed). Indeed, in this model, this invertibility is related to the nature of the operator which relates potential vorticity and streamfunction. The barotropic vorticity is the Laplacian of the barotropic streamfunction

$$q_{bt} = \nabla^2 \psi_{bt}$$

(a Poisson equation), while, for baroclinic modes, the relation between potential vorticity and streamfunction is a Helmholtz equation

$$q_{bc} = \nabla^2 \psi_{bc} - \psi_{bc}/R_d^2,$$

where R_d is the radius of deformation of the given baroclinic mode. Both types of equations are elliptical and can be inverted, provided that conditions on ψ or on velocity (its first spatial derivatives) are given at the domain boundary.

The elementary solutions of the Poisson and Helmholtz equations (that is, with Dirac distributions for potential vorticity) are the Green's functions

$$G_{bt}(x, y) = \frac{1}{2\pi} Log(r), \quad G_{bc}(x, y) = \frac{-1}{2\pi} K_0(r/R_d),$$

where $r = \sqrt{x^2 + y^2}$ and K_0 is the modified Bessel function of second kind of order zero (see also above). The solution for regular distributions of potential vorticity are therefore given by a convolution product between them and the Green's functions

$$\psi_{bt} = G_{bt} * q_{bt}$$

or

$$\psi_{bt}(x, y, t) = \int \int_{R^2} dx' dy' G_{bt}(x - x', y - y') q_{bt}(x', y', t),$$

and similarly for the baroclinic components.

Consider now a potential vorticity distribution confined to a finite domain D, as is expected for an oceanic vortex. How will the associated flow vary at large distances? If the vortex is axisymmetric, its barotropic flow will decrease as $1/r$ at large distances, while the velocity of any baroclinic mode will decrease as $K_1(r/R_d) \sim \exp(-r/R_d)$,

$$v_{bt} \sim \left[\int \int_D dx' dy' q_{bt}(x', y', t) \right] \partial_r G_{bt}(x - x', y - y')$$

$$v_{bt} \sim \left[\int \int_D dx' dy' q_{bt}(x', y', t) \right] /(2\pi r)$$

Therefore, the kinetic energy of the vortex $K \sim \int \int v^2 r dr$ will be finite if the area integral of the barotropic vorticity of the vortex is null. This can be achieved in two ways: either by having an annulus of opposite-signed vorticity around the vortex core or by having opposite-signed poles of vorticity above or below this core [108]. Obviously, if the potential vorticity distribution depends on a single spatial variable, direct integration is usually possible to obtain the associated streamfunction. A simple and well-known example is the barotropic "shielded" Gaussian vortex, which has potential vorticity

$$q_{bt}(r) = q_0(1 - r^2) \exp(-r^2) = \frac{d^2\psi_{bt}}{dr^2} + \frac{1}{r}\frac{d\psi_{bt}}{dr},$$

and a Gaussian streamfunction profile

$$\psi_{bt}(r) = (-q_0/4)\exp(-r^2).$$

Hence, potential vorticity does not solely represent the internal structure of the vortex but also the whole flow that it generates. This allows calculations of vortex stationarity, stability, and interactions.

3.2.4.3 Vortex Stationarity in the Quasi-geostrophic Model

Stationarity is expressed directly from the potential vorticity equation by canceling the time derivative (stationarity in a fixed frame of reference):

$$J(\psi_j, q_j) = 0 \quad \rightarrow q_j = F(\psi_j).$$

This is the case, for instance, of axisymmetric vortices on the f-plane. The Jacobian vanishes since ψ_j and q_j depend only on the radius r.

For stationarity in a moving frame of reference, the time derivative is replaced by the appropriate spatial derivative. For instance, stationarity in a reference frame moving at constant zonal velocity c is written as

$$J(\psi_j + cy, q_j) = 0 \quad \rightarrow q_j = F(\psi_j + cy).$$

This is the case of vortex dipoles, called modons, on the beta-plane [57, 59]. Vortices which remain stationary in a frame of reference rotating with constant rate Ω, obey the equation

$$J(\psi_j + \Omega r^2/2, q_j) = 0 \quad \rightarrow q_j = F(\psi_j + \Omega r^2/2).$$

3.2.4.4 Vortex Stability in the Quasi-geostrophic Model

We consider here the stability of circular vortices on the f-plane in a quasi-geostrophic model. The mean circular vortex is defined by $\overline{\psi}_j(r), \overline{q}_j(r)$. A normal-mode perturbation

$$\psi'_j(r, \theta, t) = \phi_j(r)\exp[il(\theta - ct)], \quad q'_j(r, \theta, t) = \xi_j(r)\exp[il(\theta - ct)]$$

is added. What are the conditions for linear instability of this perturbed vortex? The potential vorticity equation is linearized around the mean flow

$$\partial_t q'_j + J\left(\overline{\psi}_j, q'_j\right) + J\left(\psi'_j, \overline{q}_j\right) = 0,$$

which is also written as

$$(\overline{V}_j - rc)\xi_j - \frac{d\overline{q}_j}{dr}\phi_j = 0,$$

where $\overline{V}_j$ is the mean azimuthal velocity. This equation can also be written as

$$\xi_j - \frac{d\overline{q}_j}{dr} \frac{\phi_j}{\overline{V}_j - rc} = 0.$$

Multiplied by ϕ_j^*, the complex conjugate of ϕ_j and integrated over the domain area, and over layer thicknesses, this leads to

$$-E' + \sum_j H_j \frac{d\overline{q}_j}{dr} \frac{|\phi_j|^2}{\overline{V}_j - rc} = 0,$$

where E' is the perturbation energy. Since $c = c_r + ic_i$, the imaginary part of this equation is

$$c_i \sum_j H_j \frac{d\overline{q}_j}{dr} \frac{|\phi_j|^2}{(\overline{V}_j - rc_r)^2 + r^2 c_i^2} = 0.$$

To obtain positive growth rates $\sigma = lc_i$ for the perturbation (i.e., for the vortex to be unstable), a necessary condition is that $d\overline{q}_j/dr$ changes sign either in a layer or between layers. This is the Charney–Stern [31] criterion for baroclinic instability in the quasi-geostrophic model. It is a generalization of the Rayleigh [130] criterion for stratified flows.

A detailed calculation of σ when the barotropic vorticity is piecewise constant and nonlinear evolution of linearly unstable vortices can be found in [23].

3.2.5 Three-Dimensional, Boussinesq, Non-hydrostatic Models

To investigate motions which do not belong to the slow manifold (hydrostatic, balanced motions) and in particular, the breaking of inertia-gravity waves, the direct energy cascade to dissipation at small scales, intense vertical motions [164], three-dimensional Boussinesq models have been developed and used. An appropriate formulation of these equations for vortex dynamics includes potential vorticity conservation.

Usually, the 3D Boussinesq equations are written under the assumption that the averaged density distribution varies linearly along the vertical axis. We follow here the presentation of the equations given by Dritschel and Viudez and we use their notations. Density is the sum of the linear averaged density and a perturbation, and buoyancy is related to the density perturbation

$$\rho(\boldsymbol{x}, t) = \rho_0 + \rho_z z + \rho'(\boldsymbol{x}, t), \quad b = -g\rho'/\rho_0.$$

The motion is composed of a balanced part (geostrophic and hydrostatic balance) and of an imbalanced part. The balanced part is defined by

$$f\mathbf{k} \times \mathbf{u}_h = -\nabla_h \Phi/\rho_0, \quad 0 = -\partial_z \Phi/\rho_0 + b,$$

where f is the Coriolis parameter, $\mathbf{u}_h$ the horizontal velocity, and Φ is the geopotential. These equations also provide a relation between the buoyancy and the horizontal components of relative vorticity ξ and η

$$f\xi = -\partial_x b, \quad f\eta = -\partial_y b,$$

with $\xi = \partial_y w - \partial_z v$, $\eta = \partial_z u - \partial_x w$ and $\boldsymbol{\omega}(\xi, \eta, \zeta)$ with $\zeta = \partial_x v - \partial_y u$.
The imbalanced motions are described by the horizontal components of the vector

$$\mathbf{A} = \boldsymbol{\omega}/f + \nabla b/f^2,$$

which is an "ageostrophic, non-hydrostatic vorticity." Then one can define a vector velocity potential $\boldsymbol{\varphi}$ via $\mathbf{A} = \nabla^2 \boldsymbol{\varphi}$. Then $\mathbf{u}/f = -\nabla \times \boldsymbol{\varphi}$ and $D = -b/N^2 = -(1/c^2)\nabla \cdot \boldsymbol{\varphi}$.
Dimensionless potential vorticity is defined by

$$\Pi = (\boldsymbol{\omega}/f + \mathbf{k}) \cdot \nabla Z,$$

where Z is the reference height of an isopycnal defined by $Z(\mathbf{x}, t) = -g[\rho(\mathbf{x}, t)/\rho_0 - 1]/N^2 = z - D(\mathbf{x}, t)$. The potential vorticity anomaly is $\pi = \Pi - 1$. With the vector potential $\boldsymbol{\varphi} = \boldsymbol{\varphi}_h + \phi\mathbf{k}$, the following relation holds:

$$\boldsymbol{\omega}/f = \mathbf{A} - c^2\nabla D = \nabla^2 \boldsymbol{\varphi} - \nabla(\nabla \cdot \boldsymbol{\varphi}),$$

with $c = N/f$.
With these definitions, the Boussinesq equations are potential vorticity conservation (in unforced, non-dissipative conditions), relative vorticity, and imbalance equations

$$\frac{d\pi}{dt} = 0$$

$$\frac{d(\boldsymbol{\omega}/f)}{dt} = (\boldsymbol{\omega}/f) \cdot \nabla \mathbf{u} + \partial_z \mathbf{u} + fc^2\mathbf{k} \times \nabla_h D$$

$$\frac{d\mathbf{A}_h}{dt} = -f\mathbf{k} \times \mathbf{A}_h + (1 - c^2)\nabla_h w + (\boldsymbol{\omega}/f) \cdot \nabla \mathbf{u}_h + c^2\nabla_h \mathbf{u} \cdot \nabla D,$$

with $\mathbf{A}_h = \nabla^2 \boldsymbol{\varphi}_h$ and $w = dD/dt$.
Viudez and Dritschel [164] simulate the evolution of a single, baroclinic, mesoscale eddy with these equations. They observe internal gravity wave generation during the evolution of the vortex, a priori related to filamentation. With the same equations, Pallas-Sanz and Viudez [121] investigate the three-dimensional ageostrophic

motion in a mesoscale vortex dipole. For a small distance between a cyclone and an anticyclone, the vortices drift as a compact dipole and the vertical velocity pattern is octupolar. For larger separation between the vortices, the propagation speed and vertical velocities decrease and the octupolar pattern is disturbed by vortex oscillations. Dubosq and Viudez study the frontal collisions between two 3D mesoscale dipoles. The outcome can be the interchange between partners, the formation of a tripole (which is diffusion-dependent) or the squeezing of the central vortices between the outer ones.

3.3 Process Studies on Vortex Generation, Evolution, and Decay

In this section, as in the following two sections, we will show how the shallow-water models, either with PE, FG, or QG dynamics, have been used to study vortex dynamics via the analysis of individual processes.

3.3.1 Vortex Generation by Unstable Deep Ocean Jets or of Coastal Currents

The formation of vortices either from deep-ocean jets or from coastal currents has often been modeled in shallow-water or in quasi-geostrophic models. Vortex generation from these currents has been identified as resulting essentially from barotropic or baroclinic instabilities; Kelvin–Helmholtz instability, ageostrophic frontal instability, and parametric instability are other mechanisms which induce vortex shedding by such currents.

In a one-and-a-half layer quasi-geostrophic model, on the beta-plane, Flierl et al. [58] evidence a variety of nonlinear regimes of a barotropically unstable Gaussian jet depending on the wavelength and beta-effect: dipoles form for long waves at low beta, staggered vortex streets for intermediate wavelengths and cat's eyes for short waves. At higher values of beta, multi-stage instability is observed where harmonics develop and interact under the form of meanders, accompanied by Rossby wave radiation.

In a multi-layer quasi-geostrophic model, barotropic and baroclinic jet instability leads to meanders which amplify to form eddies [72, 73]. Eddy detachment is assisted by beta-effect which then restores the zonal mean flow. Flierl et al. [56] determine the nonlinear regimes of a mixed barotropically–baroclinically unstable jet and analyze the similarity with the two-dimensional case [58]. Meacham [107] studies the stability of a baroclinic jet with piecewise constant potential vorticity; he finds that the nonlinear regimes of vortex formation are related to the linear stability properties of the jet and that the most realistic nonlinear jet evolutions are obtained for a single potential vorticity front in the upper and lower layers.

In a multi-layer shallow-water model, Boss et al. [16] show that several types of modes can develop on an unstable outcropping front in a two-layer SW model:

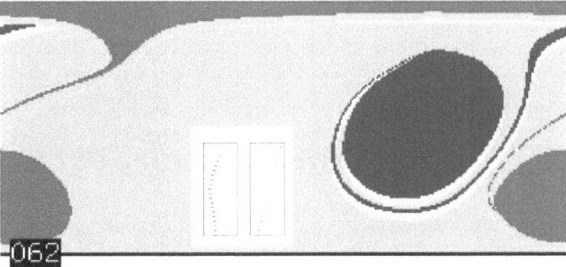

Fig. 3.9 Baroclinic dipole formation and ejection from an unstable coastal current; from Cherubin et al. [34]

Kelvin-like modes (those previously observed for frontal instability) and Rossby-like modes (related to baroclinic instability). Baey et al. [8] show that the instability of identical jets is stronger in the SW model than in the quasi-geostrophic model and that anticyclones seem to appear more often and are larger than cyclones in the former model.

Chérubin et al. [33] investigate the linear stability of a two-dimensional coastal current composed of two adjacent uniform vorticity strips and found evidence of dipole formation when the instability is triggered by a canyon. In contrast, stable flows (made of a single vorticity strip) shed filaments near deep canyons. Capet and Carton [22] study the nonlinear regimes of the same QG flow over a flat bottom or over a topographic shelf. They find that the critical parameter for water export offshore is the distance from the coast where the phase speed of the waves equals the mean flow velocity. Chérubin et al. [34] study the baroclinic instability of the same flow over a continental slope with application to the Mediterranean Water (MW) undercurrents: vortex dipoles similar to the dipoles of MW can form for long waves when layerwise PV amplitudes are comparable but of opposite sign (see Fig. 3.9). This confirms the Stern et al. [151] results of laboratory experiments and primitive-equation modeling which show that dipoles can form from unstable coastal currents as in two-dimensional flows.

3.3.2 Vortex Generation by Currents Encountering a Topographic Obstacle

The interaction of a flow with an isolated seamount is a longstanding problem in oceanography, and in a homogeneous fluid the classical solution of the Taylor column is well known. When the flow varies with time, when the fluid is stratified, or when the topographic obstacle is more complex, several studies have provided essential results on vortex generation.

Verron [163] addressed the formation of vortices by a time-varying barotropic flow over an isolated seamount. He found that vortices are shed by topographic obstacles of intermediate height. Small topographies do not trap particles above them (they are advected by the flow). Tall topographies do not release significant amounts of water. The conditions under which vortices can be shed by a seamount in a uniform flow are given in Huppert [70] and Huppert and Bryan [71].

3.3.3 Vortex Generation by Currents Changing Direction

Many oceanic eddies are formed near capes where coastal currents change direction. Ou and De Ruijter [118] relate the flow separation from the coast to the outcropping of the current at the coast as it veers around the cape. Another mechanism, based on vorticity generation in the frictional boundary layer, is proposed for the formation of submesoscale coherent vortices, when the current turns around a cape [45]. Klinger [80–82] finds a condition on the curvature of the coast to obtain flow separation, and in the case of a sharp angle, he observes the formation of a gyre at the cape for a 45° angle and eddy detachment at a 90° angle.

Nof and Pichevin [114] and Pichevin and Nof [125, 126] propose a theory for currents changing direction, e.g., as they exit from straits or veer around capes. In this case, linear momentum is not conserved in all directions (see Fig. 3.10a). Indeed an integration of the SW equations in flux form over the domain ABCDEFA leads to

$$\int_C^D [hu^2 + g'h^2/2 - f\psi]\, dy = 0$$

via the definition of a transport streamfunction ψ and the Stokes' theorem. With the geostrophic balance

$$f\psi = g'h^2/2 - \beta \int_y^L \psi dy$$

the previous equation becomes

$$\int_0^L hu^2 dy + \beta \int_0^L [\int_y^L \psi dy]dy = 0,$$

which cannot be satisfied since both terms are positive.

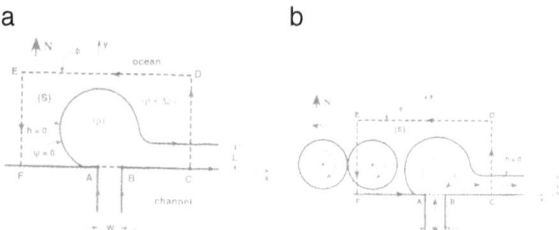

Fig. 3.10 (a) *Top*: sketch of the current exiting from the strait without vortex formation; (b) *bottom*: same as (a) but now with vortex generation; from Pichevin and Nof [126]

The equilibrium is then reached in time by periodic formation of vortices which exit the domain in the opposite direction to the mean flow (see Fig. 3.10b). By defining a time-averaged transport streamfunction $\tilde{\psi}$ (over a period T of vortex shedding), the balance then becomes

$$\int_C^D [hu^2 + g'h^2/2 - f\psi]\, dy = \int_0^T \int_F^E [hu^2 + g'h^2/2]\, dy\, dt - \int_F^E f\tilde{\psi} dy.$$

The flow force exerted on the domain by the water exiting from its right is balanced by eddies shed on the left.

Numerical experiments with a PE model indeed show that vortices periodically grow and detach from the current, when this current changes direction (see Fig. 3.11). This can explain the formation of meddies at Cape Saint Vincent, of Agulhas rings south of Africa, of Loop Current eddies in the Gulf of Mexico, of teddies (Indonesian Throughflow eddies), etc. (see Sect. 3.1.2).

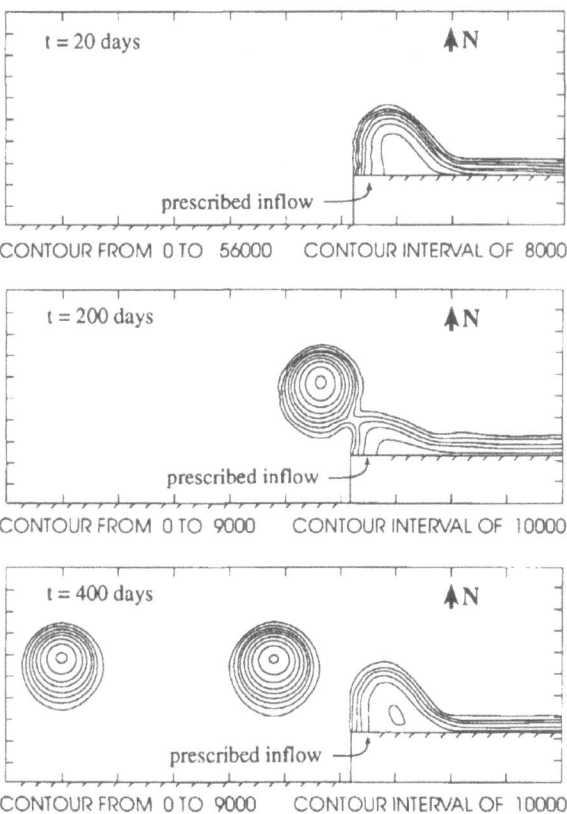

Fig. 3.11 Result of PE model simulation; from Pichevin and Nof [126]

3.3.4 Beta-Drift of Vortices

First, let us recall the basic idea behind the motion of vortices on the beta-plane. Consider an isolated lens eddy (see, for instance, [111] or [79]): since f varies with latitude, the southward Coriolis force acting on the northern side of an anti-cyclone will be stronger than the opposite force acting on its southern side (in the northern hemisphere). Hence circular lens eddies cannot remain motionless on the beta-plane. To balance this excess of meridional force, a northward Coriolis force associated with a westward motion is necessary. For a cyclone, the converse rea-soning leads to an eastward motion which is not observed. Why? Because cyclones are not isolated mass anomalies (the isopycnals do not pinch off). Therefore, they entrain the surrounding fluid and the motion of this fluid must be taken into account. The surrounding fluid advected northward (resp. southward) by the vortex flow will lose (resp. gain) relative vorticity, creating a dipolar vorticity anomaly which will push the cyclone westward. This mechanism is responsible in part for the creation of the so-called beta-gyres (see Fig. 3.12).

In summary, on the beta-plane, both a deformation and a global motion of the vortex will occur. Now we provide a short summary of the mathematics of the problem, essentially for two-dimensional vortices, with piecewise-constant vorticity distri-butions. These mathematics describe the first stage of the beta-drift in which the influence of the far-field of the Rossby wave wake is not important. In the ocean, his effect becomes dominant after a few weeks. This wake drains energy from the vortex and the mathematical model of its interaction with the vortex at late stages is still an open problem.

For a piecewise-constant vortex, assuming a weak beta-effect relative to the vor-tex strength (on order ϵ), Sutyrin and Flierl have shown that one part of the beta-gyre potential vorticity is due to the advection of the planetary vorticity by the azimuthal vortex flow. The PV anomaly is then of order ϵ and its normalized amplitude is

$$q = r[\sin(\theta - \Omega t) - \sin(\theta)] = \nabla^2 \phi - \gamma^2 \phi,$$

where Ω is the rotation rate of the mean flow and $\gamma = 1/R_d$. The other part is due to the deformation of the vortex contour due to its advection by the first part of the

Fig. 3.12 Early development of beta-gyres on a Rankine vortex in a 1-1/2 QG model, with $R = R_d$ and $\beta R_d / q_{max} = 0.04$

beta-gyres. Assuming a mode 1 deformation and a single vortex contour, one has
the following time-evolution equation for the vortex contour $r = 1 + \eta(t)\exp(i\theta)$:

$$dn/dt - i[\Omega(r) + \frac{\Delta}{r}G_1(r/1)]\eta = i\frac{\phi}{r} - u - iv,$$

with u and v the drift velocities, G_1 the Green's function for the Helmholtz problem with $\exp(i\theta)$ dependence, and Δ is the PV jump across the vortex boundary.
Choosing $\Omega(1) = 1$, one obtains the following drift velocity (in normalized form):

$$u + iv = \frac{-1}{\gamma^2} + \int G_1(r/1)\ \exp(i\Omega(r)t)\ r^2\ dr.$$

This theory does not model the far field of the wave separately. The nonlinear evolution of the vortex will induce a transient mode 2 deformation in the vortex contour
so that temporary tripolar states can be observed [153]. This will create cusps in the
trajectories, where these tripoles stagnate and tumble. Lam and Dritschel [83] investigate numerically the influence of the vortex amplitude and radius on its beta-drift
in the same framework. They observe that the zonal speed of a vortex increases with
its size. Large and weak vortices are often deformed, elliptically or into tripoles.
Furthermore, strong gradients of vorticity appear around and behind the vortex: the
gradient circling around the vortex forms a trapped zone which shrinks with time,
while the trailing front extends behind the vortex. The interaction of these vortex
sheets with the vortex still needs mathematical modeling.

3.3.5 Interaction Between a Vortex and a Vorticity Front or a Narrow Jet

Bell [9] investigates the interaction between a point vortex and a PV front in a 1-1/2
layer QG model. The asymptotic theory of weak interaction (small deviations of the
PV front) leads to the result that a spreading packet of PV front waves will form in
the lee of the vortex, thus transferring momentum from the vortex to the front, and
that the meander close to the vortex will induce a transverse motion on the vortex
(toward or away from the front). Stern [150] extends this work to a finite-area vortex
in a 2D flow and finds that the drift velocity of the vortex along the front scales with
the square root of the vorticity products (of the vortex and of the shear flow). He
observes wrapping of the front around the vortex. Bell and Pratt [10] consider the
case of an unstable jet interacting with a vortex in QG models with a single active
layer. In the 2D case, the jet breaks up in eddies while in the 1-1/2 layer case, the jet
is stable and long waves develop on the front and advect the vortex in the opposite
direction to the 2D case.

Vandermeirsch et al. [159, 160] investigate the conditions under which an eddy
can cross a zonal jet, with application to meddies and to the Azores Current. They
find that a critical point of the flow must exist on the jet axis to allow this crossing

and this condition can be expressed both in QG and SW models. They further address the case of an unstable surface-intensified jet in a two-layer model and show that

(a) a baroclinic dipole is formed south of the jet (for an eastward jet interacting with an anticyclone coming from the North) and
(b) the meanders created by vortex-jet interaction clearly differ in length from those of the baroclinic instability of the jet.

Therefore, the interaction is identifiable, even for a deep vortex. Such an interaction was indeed observed with these characteristics in the Azores region during the Semaphore 1993 experiment at sea [158].

3.3.6 Vortex Decay by Erosion Over Topography

The interaction of a vortex with a seamount has been often studied, bearing in mind its application to meddies interacting with Ampere Seamount or Agulhas rings with the Vema seamount. Van Geffen and Davies [161] model the collision of a monopolar vortex on a seamount on the beta-plane in a 2D flow. Large seamounts in the southern hemisphere can deflect the vortex northward or back to the southeast while in the northern hemisphere, the monopole will be strongly deformed and its further trajectory complex. Cenedese [25] performs laboratory experiments and evidences peeling off of the vortex by topography and substantial deflection as for meddies encountering seamounts. Herbette et al. [66, 67] model the interaction of a surface vortex with a tall isolated seamount, with application to the Agulhas rings and the Vema seamount. On the f-plane, they find that the surface anticyclone is eroded and may split, in the shear and strain flow created by the topographic vortices in the lower layer. Sensitivity of these behaviors to physical parameters is assessed. On the beta-plane, these effects are even more complicated due to the presence of additional eddies created by the anticyclone propagation. In the case of a tall isolated seamount, the most noticeable effect is the circulation and shear created by the anticyclonic topographic vortex and the incident vortex trajectory can be explained by its position relative to a flow separatrix [152].

3.4 Conclusions

This review of oceanic vortices has deliberately neglected the aspects of mutual vortex interactions and vortices in oceanic turbulence, which have been described in McWilliams [100] and in Carton [23]. These aspects are nevertheless important. The first part of the present review has illustrated the diversity of oceanic eddies and of their evolutions (formation mechanisms, interactions with neighboring currents or with topography, decay). Though surface-intensified eddies have received more attention earlier, intrathermocline eddies (such as meddies) have been sampled, described, and analyzed in great detail in the past 20 years, due to progress in technology (in particular, for acoustically tracked floats). Nevertheless, for deep eddies, the generation mechanisms in the presence of fluctuating currents and over complex topography are not completely elucidated.

Many measurements at sea are still needed to provide a detailed description of oceanic eddies, in particular in the coastal regions and near the outlets of marginal seas. The global network for ocean monitoring, based on profiling floats, on hydrological and current-meter measurements, and on satellite observations, will certainly bring interesting information in that respect, but it needs to be densified in the coastal regions. New tools such as seismic imaging of water masses may provide a high vertical and horizontal resolution and spatial continuity in the measurement of water masses. The relative influence of beta-effect, topography (or continental boundaries), and barotropic or vertically sheared currents over the propagation of oceanic vortices also needs further assessment. Little work has been performed on the decay of vortices via ventilation. The relation of eddy structure to fine-scale mixing is a current subject of investigation.

Vortex interaction, both mutual and with surrounding currents or topography, has proved an important source for smaller-scale motions (submesoscale filaments, for instance, see [53]). Recent work [88, 84, 85] shows that these filaments are the sites of intense vertical motion near the sea surface and below, effectively bringing nutrients in the euphotic layer, for instance, and contributing more efficiently to the biological pump than the vortex cores (as traditionally believed). This research field is certainly essential for an improved understanding of upper ocean turbulence and biological activity.

More generally, a research path of central importance for the years to come is the interactions between motions of notably different spatial and temporal scales. The relations between submesoscale, mesoscale, synoptic, basin, and planetary-scale motions are a completely open field, to which, undoubtedly, the past work on vortex dynamics will contribute.

Acknowledgments The author is grateful to the scientific committee and the local organizers of the Summer school for the excellent scientific exchanges and for the hospitality at Valle d'Aosta. Sincere thanks are due to an anonymous referee and to Drs Bernard Le Cann and Alain Serpette for their careful reading of this text and for their fine suggestions.
This work was supported in part by the INTAS contract "Vortex Dynamics" (project 7297, collaborative call with Airbus); it is a contribution to the ERG "Regular and chaotic hydrodynamics."

References

1. Adem, J.: A series solution for the barotropic vorticity equation and its application in the study of atmospheric vortices, Tellus, **8**, 364 (1956).
2. Arhan, M., Colin de Verdiere, A., Memery, L.: The eastern boundary of the subtropical North Atlantic, J. Phys. Oceanogr., **24**, 1295 (1994).

3. Arhan, M., Mercier, H., Lutjeharms, J.R.E.: The disparate evolution of three Agulhas rings in the South Atlantic Ocean, J. Geophys. Res., **104** (C9), 20987 (1999).

4. Armi, L., Hedstrom, K.: An experimental study of homogeneous lenses in a stratified rotating fluid, J. Fluid Mech., **191**, 535 (1988).

5. Armi, L., Stommel, H.: Four views of a portion of the North Atlantic subtropical gyre, J. Phys. Oceanogr., **13**, 828 (1983).

6. Armi, L., Zenk, W.: Large lenses of highly saline Mediterranean water, J. Phys. Oceanogr., **14**, 1560 (1984).

7. Armi, L., Hebert, D., Oakey, N., Price, J.F., Richardson, P.L., Rossby, T.M., Ruddick, B.: Two years in the life of a Mediterranean Salt Lens, J. Phys. Oceanogr., **19**, 354 (1989).

8. Baey, J.M., Riviere, P., Carton, X.: Ocean jet instability: a model comparison. In European Series in Applied and Industrial Mathematics: Proceedings, Vol. 7, SMAI, Paris, pp. 12–23 (1999).

9. Bell, G.I.: Interaction between vortices and waves in a simple model of geophysical flow, Phys. Fluids A, **2**, 575 (1990).

10. Bell, G.I., Pratt, L.J.: The interaction of an eddy with an unstable jet, J. Phys. Oceanogr., **22**, 1229 (1992).

11. Benilov, E.S.: Large-amplitude geostrophic dynamics: the two-layer model, Geophys. Astrophys. Fluid Dyn., **66**, 67 (1992).

12. Benilov, E.S.: Dynamics of large-amplitude geostrophic flows: the case of 'strong' beta-effect, J. Fluid Mech., **262**, 157 (1994).

13. Benilov, E.S., Cushman-Roisin, B.: On the stability of two-layered large-amplitude geostrophic flows with thin upper layer, Geophys. Astrophys. Fluid Dyn., **76**, 29 (1994).

14. Benilov, E.S., Reznik, G.M.: The complete classification of large-amplitude geostrophic flows in a two-layer fluid, Geophys. Astrophys. Fluid Dyn., **82**, 1 (1996).

15. Bolin, B.: Numerical forecasting with the barotropic model, Tellus, **7**, 27 (1955).

16. Boss, E., Paldor, N., Thomson, L.: Stability of a potential vorticity front: from quasi-geostrophy to shallow water, J. Fluid Mech., **315**, 65 (1996).

17. Boudra, D.B., Chassignet, E.P.: Dynamics of the Agulhas retroflection and ring formation in a numerical model, J. Phys. Oceanogr., **18**, 280 (1988).

18. Boudra, D.B., De Ruijter, W.P.M.: The wind-driven circulation of the South Atlantic-Indian Ocean. II: Experiments using a multi-layer numerical model, Deep-Sea Res., **33**, 447 (1986).

19. Bower, A., Armi, L., Ambar, I.: Direct evidence of meddy formation off the southwestern coast of Portugal, Deep-Sea Res., **42**, 1621 (1995).

20. Bower, A., Armi, L., Ambar, I.: Lagrangian observations of meddy formation during a Mediterranean undercurrent seeding experiment, J. Phys. Oceanogr., **27**, 2545 (1997).

21. Bretherton, F.P.: Critical layer instability in baroclinic flows, Q. J. Roy. Met. Soc., **92**, 325 (1966).

22. Capet, X., Carton, X.: Nonlinear regimes of baroclinic boundary currents, J. Phys. Oceanogr., **34**, 1400 (2004).

23. Carton, X.: Hydrodynamical modeling of oceanic vortices, Surveys Geophys., **22**, 3, 179 (2001).

24. Carton, X., Chérubin, L., Paillet, J., Morel, Y., Serpette, A., Le Cann, B.: Meddy coupling with a deep cyclone in the Gulf of Cadiz, J. Mar. Syst., **32**, 13 (2002).

25. Cenedese, C.: Laboratory experiments on mesoscale vortices colliding with a seamount. J. Geophys. Res. C, **107**, 3053 (2002).

26. Chao, S.Y., Kao, T.W.: Frontal instabilities of baroclinic ocean currents, with applications to the gulf stream, J. Phys. Oceanogr., **17**, 792 (1987).

27. Chapman, R., Nof, D.: The sinking of warm-core rings, J. Phys. Oceanogr., **18**, 565 (1988)

28. Charney, J.: Dynamics of long waves in a baroclinic westerly current, J. Meteor., **4**, 135 (1947).

29. Charney, J.: On the scale of atmospheric motions, Geofys. Publikasjoner, **17**, 1 (1948).

30. Charney, J.: The use of the primitive equations of motion in numerical prediction, Tellus, **7**, 22 (1955).
31. Charney, J.G., Stern, M.E.: On the stability of internal baroclinic jets in a rotating atmosphere, J. Atmos. Sci., **19**, 159 (1962).
32. Chassignet, E.P., Boudra, D.B.: Dynamics of Agulhas retroflection and ring formation in a numerical model, J. Phys. Oceanogr., **18**, 304 (1988).
33. Chérubin, L.M., Carton, X., Dritschel, D.G.: Vortex expulsion by a zonal coastal jet on a transverse canyon. In Proceedings of the Second International Workshop on Vortex Flows, Vol. 1, SMAI, Paris, pp. 481–501 (1996).
34. Chérubin, L.M., Carton, X., Dritschel, D.G.: Baroclinic instability of boundary currents over a sloping bottom in a quasi-geostrophic model. J. Phys. Oceanogr., **37**, 1661 (2007).
35. Chérubin, L.M., Carton, X., Paillet, J., Morel, Y., Serpette, A.: Instability of the Mediterranean water undercurrents southwest of Portugal: effects of baroclinicity and of topography, Oceanologica Acta, **23**, 551 (2000).
36. Chérubin, L.M., Serra, N., Ambar, I.: Low frequency variability of the Mediterranean undercurrent downstream of Portimão Canyon, J. Geophys. Res. C, **108**, 10.1029/2001JC001229 (2003).
37. Creswell, G.: The coalescence of two East Australian current warm-core eddies, Science, **215**, 161 (1982).
38. Cronin, M.: Eddy-mean flow interaction in the Gulf stream at 68°W: Part II. Eddy forcing on the time-mean flow, J. Phys. Oceanogr., **26**, 2132 (1996).
39. Cronin, M., Watts, R.D.: Eddy-mean flow interaction in the Gulf stream at 68°W: Part I. Eddy energetics, J. Phys. Oceanogr., **26**, 2107 (1996).
40. Cushman-Roisin, B.: Introduction to Geophysical Fluid Dynamics, Prentice-Hall, New Jersey, 320pp. (1994).
41. Cushman-Roisin, B.: Frontal geostrophic dynamics, J. Phys. Oceanogr., **16**, 132 (1986).
42. Cushman-Roisin, B., Tang, B.: Geostrophic turbulence and emergence of eddies beyond the radius of deformation, J. Phys. Oceanogr., **20**, 97 (1990).
43. Cushman-Roisin, B., Sutyrin, G.G., Tang, B.: Two-layer geostrophic dynamics. Part I: Governing equations, J. Phys. Oceanogr., **22**, 117 (1992).
44. Danielsen, E.F.: In defense of Ertel's potential vorticity and its general applicability as a meteorological tracer, J. Atmos. Sci., **47**, 2013 (1990).
45. D'Asaro, E.: Generation of submesoscale vortices: a new mechanism, J. Geophys. Res. C, **93**, 6685 (1988).
46. De Ruijter, W.P.M.: Asymptotic analysis of the Agulhas and Brazil current systems, J. Phys. Oceanogr., **12**, 361 (1982).
47. De Ruijter, W.P.M., Boudra, D.B.: The wind-driven circulation in the South Atlantic-Indian Ocean–I. Numerical experiments in a one layer model, Deep-Sea Res., **32**, 557 (1985).
48. Dijkstra, H.A., De Ruijter, W.P.M.: Barotropic instabilities of the Agulhas Current system and their relation to ring formation, J. Mar. Res., **59**, 517 (2001).
49. Duncombe-Rae, C.M.: Agulhas retrollection rings in the South Atlantic Ocean: an overview, S. Afr. J. Mar. Sci., **11**, 327 (1991).
50. Evans, R., Baker, k.S., Brown, O., Smith, R.: Chronology of warm-core ring 82-B, J. Geophys. Res., **90**, 8803 (1985).
51. Ezer, T.: On the interaction between the Gulf stream and the New England seamount chain, J. Phys. Oceanogr., **24**, 191 (1994).
52. Fedorov, K.N., Ginsburg, A.I.: Mushroom-like currents (vortex dipoles): one of the most widespread forms of non-stationary coherent motions in the ocean. In: Nihoul, J.C.J., Jamart, B.M. (eds.) Mesoscale/Synoptic Coherent Structures in Geophysical Turbulence, Vol. 50, Elsevier Oceanographic Series, Amsterdam, pp. 1–14 (1989).
53. Flament, P., Armi, L., Washburn, L.: The evolving structure of an upwelling filament, J. Geophys. Res., **90**, 11, 765 (1985).

54. Flament, P., Lupmkin, R., Tournadre, J., Armi, L.: Vortex pairing in an anticylonic shear flow: discrete subharmonics of one pendulum day, J. Fluid Mech., **440**, 401 (2001).
55. Flierl, G.R.: Isolated eddy models in geophysics, Ann. Rev. Fluid Mech., **19**, 493 (1987).
56. Flierl, G.R., Carton, X.J., Messager, C.: Vortex formation by unstable oceanic jets. In European Series in Applied and Industrial Mathematics: Proceedings, Vol. 7, SMAI, Paris, pp. 137–150 (1999).
57. Flierl, G.R., Larichev, V.D., Mc Williams, J.C., Reznik, G.M.: The dynamics of baroclinic and barotropic solitary eddies, Dyn. Atmos. Oceans, **5**, 1 (1980).
58. Flierl, G.R., Malanotte-Rizzoli, P., Zabusky, N.J.: Nonlinear waves and coherent vortex structures in barotropic beta plane jets, J. Phys. Oceanogr., **17**, 1408 (1987).
59. Flierl, G.R., Stern, M.E., Whitehead, J.A.: The physical significance of modons: laboratory experiments and general integral constraints, Dyn. Atmos. Oceans, **7**, 233 (1983).
60. Garzoli, S.L., Ffield, A., Johns, W.E., Yao, Q.: North Brazil Current retroflection and transport, J. Geophys. Res. Oceans, **109**, C1 (2004).
61. Garzoli, S.L., Yao, Q., Ffield, A.: *Interhemispheric Water Exchange in the Atlantic Ocean.* (eds. Goni, G., Malanotte-Rizzoli, P.), Elsevier Oceanographic Series, Amsterdam, pp 357–374 (2003).
62. Gill, A.E.: Homogeneous intrusions in a rotating stratified fluid, J. Fluid Mech., **103**, 275 (1981).
63. Gill, A.E., Schumann, E.H.: Topographically induced changes in the structure of an inertial coastal jet: application to the Agulhas Current, J. Phys. Oceanogr., **9**, 975 (1979).
64. Haynes, P.H., McIntyre, M.E.: On the representation of Rossby-wave critical layers and wave breaking in zonally truncated models, J. Atmos. Sci., **44**, 828 (1987).
65. Haynes, P.H., McIntyre, M.E.: On the conservation and impermeability theorems for potential vorticity, J. Atmos. Sci., **47**, 2021 (1990).
66. Herbette, S., Morel, Y., Arhan, M.: Erosion of a surface vortex by a seamount, J. Phys. Oceanogr., **33**, 1664 (2003).
67. Herbette, S., Morel, Y., Arhan, M.: Erosion of a surface vortex by a seamount on the beta plane, J. Phys. Oceanogr., **35**, 2012 (2005).
68. Hoskins, B.J.: The geostrophic momentum approximation and the semi-geostrophic equations, J. Atmos. Sci., **32**, 233 (1975).
69. Hoskins, B.J., McIntyre, M.E., Robertson, A.: On the use and significance of isentropic potential vorticity maps, Q. J. Roy. Met. Soc., **111**, 887 (1985).
70. Huppert, H.E.: Some remarks on the initiation of inertial Taylor columns, J. Fluid Mech., **67**, 397 (1975).
71. Huppert, H.E., Bryan, K.: Topographically generated eddies, Deep-Sea Res., **23**, 655 (1976).
72. Ikeda, M.: Meanders and detached eddies of a strong eastward-flowing jet using a two-layer quasi-geostrophic model, J. Phys. Oceanogr., **11**, 525 (1981).
73. Ikeda, M., Apel, J.R.: Mesoscale eddies detached from spatially growing meanders in an eastward flowing oceanic jet, J. Phys. Oceanogr., **11**, 1638 (1981).
74. Joyce, T.M., Backlus, R., Baker, K., Blackwelder, P., Brown, O., Cowles, T., Evans, R., Fryxell, G., Mountain, D., Olson, D., Shlitz, R., Schmitt, R., Smith, P., Smith, R., Wiebe, P.: Rapid evolution of a Gulf Stream warm-core ring, Nature, **308**, 837 (1984).
75. Joyce, T.M., Stalcup, M.C.: Wintertime convection in a Gulf Stream warm core ring, J. Phys. Oceanogr., **15**, 1032 (1985).
76. Kamenkovich, V.M., Koshlyakov, M.N., Monin, A.S.: Synoptic Eddies in the Ocean, EFM, D. Reidel Publ. Company, Dordrecht, 433pp (1986).
77. Karsten, R.H., Swaters, G.E.: A unified asymptotic derivation of two-layer, frontal geostrophic models including planetary sphericity and variable topography, Phys. Fluids, **11**, 2583 (1999).
78. Kennan, S.C., Flament, P.J.: Observations of a tropical instability vortex, J. Phys. Oceanogr., **30**, 2277 (2000).

79. Killworth, P.D.: Long-wave instability of an isolated front, Geophys. Astrophys. Fluid Dyn., **25**, 235 (1983).
80. Klinger, B.A.: Gyre formation at a corner by rotating barotropic coastal flows along a slope, Dyn. Atmos. Oceans, **19**, 27 (1993).
81. Klinger, B.A.: Inviscid current separation from rounded capes, J. Phys. Oceanogr., **24**, 1805 (1994a).
82. Klinger, B.A.: Baroclinic eddy generation at a sharp corner in a rotating system, J. Geophys. Res. C6, **99**, 12515 (1994b).
83. Lam, J.S-L., Dritchel, D.G.: On the beta-drift of an initially circular vortex patch, J. Fluid Mech., **436**, 107 (2001).
84. Lapeyre, G., Klein, P.: Impact of the small-scale elongated filaments on the oceanic vertical pump, J. Mar. Res., **64**, 835 (2006).
85. Lapeyre, G., Klein, P.: Dynamics of the upper oceanic layers in terms of surface quasigeostrophy theory, J. Phys. Oceanogr., **36**, 165 (2006).
86. Legg, S.A.: Open Ocean Deep Convection : The Spreading Phase. PhD Thesis, Imperial College, University of London, London (1992).
87. Leith, C.E.: Nonlinear normal mode initialization and quasi-geostrophic theory, J. Atmos. Sci., **37**, 958 (1980).
88. Levy, M., Klein, P., Treguier, A.M.: Impacts of sub-mesoscale physics on phytoplankton production and subduction, J. Mar. Res., **59**, 535 (2001).
89. Lorenz, E.N.: Attractor sets and quasi-geostrophic equilibrium, J. Atmos. Sci., **37**, 1685 (1980).
90. Lutjeharms, J.R.E.: The Agulhas Current, 1st edn. Springer, Berlin, pp. 113–207 (2006).
91. Lutjeharms, J.R.E., van Ballegooyen, R.C.: Topographic control in the Agulhas Current system, Deep-Sea Res., **31**, 1321 (1984).
92. Lutjeharms, J.R.E., van Ballegooyen, R.C.: The Retroflection of the Agulhas Current, J. Phys. Oceanogr., **18**, 1570 (1988).
93. Lutjeharms, J.R.E., Penven, P., Roy, C.: Modelling the shear edge eddies of the southern Agulhas Current, Cont. Shelf Res., **23**, 1099 (2003).
94. Ma, H.: The dynamics of North Brazil Current retroflection eddies, J. Mar. Res., **54**, 35 (1996).
95. Mazé, J.P., Arhan, M., Mercier, H.: Volume budget of the eastern boundary layer off the Iberian Peninsula, Deep-Sea Res., **44**, 1543 (1997).
96. McIntyre, M.E., Norton, W.A.: Dissipative wave-mean interactions and the transport of vorticity or potential vorticity, J. Fluid Mech., **212**, 403 (1990).
97. McIntyre, M.E., Norton, W.A.: Potential vorticity inversion on a hemisphere, J. Atmos. Sci, **57**, 1214 (2000).
98. McIntyre, M.E.: Spontaneous Imbalance and hybrid vortex-gravity structures, J. Atmos. Sci., **66** (5), 1315–1326 (2009).
99. Mc Williams, J.C.: Submesoscale, coherent vortices in the ocean, Rev. Geophys., **23**, 165 (1985).
100. Mc Williams, J.C.: Geostrophic vortices. In: Nonlinear Topics in Ocean Physics. Proceedings of the International School of Physics "Enrico Fermi", Course CIX, North Holland, New York, pp. 5–50 (1991).
101. Mc Williams, J.C.: Diagnostic force balance and its limits. In: Velasco Fuentes, O.U., Sheinbaum, J., Ochoa, J. (eds.) Nonlinear Processes in Geophysical Fluid Dynamics, Vol. 287 Kluwer Acadamic Publishers, Dordrecht (2003).
102. Mc Williams, J.C., Flierl, G.R.: On the evolution of isolated, non-linear vortices, J. Phys. Oceanogr., **9**, 1155 (1979).
103. Mc Williams, J.C., Gent, P.R.: The evolution of sub-mesoscale, coherent vortices on the beta-plane, Geophys. Astrophys. Fluid Dyn., **35**, 235 (1986).
104. Mc Williams, J.C., Gent, P.R.: Intermediate models of planetary circulations in the atmosphere and ocean, J. Atmos. Sci., **37**, 1657 (1980).

105. Mc Williams, J.C., Yavneh, I.: Fluctuation growth and instability associated with a singularity of the balance equations, Phys. Fluids, **10**, 2587 (1998).
106. Mc Williams, J.C., Gent, P.R., Norton, N.: The evolution of balanced, low-mode vortices on the beta-plane, J. Phys. Oceanogr., **16**, 838 (1986).
107. Meacham, S.P.: Meander evolution on piecewise-uniform, quasi-geostrophic jets, J. Phys. Oceanogr., **21**, 1139 (1991).
108. Morel, Y.: Modélisation des processus océaniques à moyenne échelle. Habilitation à diriger les recherches, Université de Bretagne Occidentale, 47pp. (2005).
109. Morel, Y., Mc Williams, J.C.: Effects of isopycnal and diapycnal mixing on the stability of ocean currents, J. Phys. Oceanogr., **31**, 2280 (2001).
110. Morel, Y., Darr, D.S., Talandier, C.: Possible sources driving the potential vorticity structure and long-wave instability of coastal upwelling and downwelling currents, J. Phys. Oceanogr., **36**, 875 (2006).
111. Nof, D.: On the beta-induced movement of isolated baroclinic eddies, J. Phys. Oceanogr., **11**, 1662 (1981).
112. Nof, D.: On the migration of isolated eddies with application to Gulf Stream rings, J. Mar. Res., **41**, 399 (1983).
113. Nof, D.: The momentum imbalance paradox revisited, J. Phys. Oceanogr., **35**, 1928 (2005).
114. Nof, D., Pichevin, T.: The retroflection paradox, J. Phys. Oceanogr., **26**, 2344 (1996).
115. Oey, L.Y.: A model of Gulf Stream frontal instabilities, meanders and eddies along the continental slope, J. Phys. Oceanogr., **18**, 211 (1988).
116. Olson, D.B.: The physical oceanography of two rings observed by cyclonic ring experiment, Part II: dynamics, J. Phys. Oceanogr., **10**, 514 (1980).
117. Olson, D.B., Schmitt, R.W., Kennelly, M., Joyce, T.M.: A two-layer diagnostic model of the long term physical evolution of warm core ring 82B, J. Geophys. Res., **90**, C5, 8813 (1985).
118. Ou, H.W., De Ruijter, W.P.M.: Separation of an inertial boundary current from a curved coastline, J. Phys. Oceanogr., **16**, 280 (1986).
119. Paillet, J., LeCann, B., Carton, X., Morel, Y., Serpette, A.: Dynamics and evolution of a northern meddy, J. Phys. Oceanogr., **32**, 55 (2002).
120. Paillet, J., LeCann, B., Serpette, A., Morel, Y., Carton, X.: Real-time tracking of a northern meddy in 1997–98, Geophys. Res. Lett., **26**, 1877 (1999).
121. Pallas-Sanz, E., Viudez, A.: Three-dimensional ageostrophic motion in mesoscale vortex dipoles, J. Phys. Oceanogr., **37**, 84 (2007).
122. Pavec, M., Carton, X., Herbette, S., Roullet, G., Mariette, V.: Instability of a coastal jet in a two-layer model ; application to the Ushant front. Proceedings of the 18th Congrès Francais de Mécanique, Grenoble (2007).
123. Pedlosky, J.: Geophysical Fluid Dynamics. Springer Verlag, New York, 624 pp (1987).
124. Phillips, N.A.: On the problem of initial data for the primitive equations, Tellus, **12**, 121 (1960).
125. Pichevin, T., Nof, D.: The eddy cannon, Deep-Sea Res., **43**, 1475 (1996).
126. Pichevin, T., Nof, D.: The momentum imbalance paradox, Tellus, **49**, 298 (1997).
127. Pichevin, T., Nof, D., Lutjeharms, J.R.E.: Why are there. Agulhas rings? J. Phys. Oceanogr., **29**, 693 (1999).
128. Pingree, R., Le Cann, B.: Three anticyclonic Slope Water Ocenic eDDIES (SWODDIES) in the southern Bay of Biscay in 1990, Deep-Sea Res., **39**, 1147 (1992).
129. Pingree, R., Le Cann, B.: A shallow meddy (a smeddy): from the secondary mediterranean salinity maximum, J. Geophys. Res. C, **98**, 20169 (1993).
130. Rayleigh, L.: On the stability or instability of certain fluid motions, Proc. Lond. Math Soc., **11**, 57 (1880).
131. Richardson, P.L.: Tracking ocean eddies, Am. Sci., **81**, 261 (1993).
132. Richardson, P.L., Tychensky, A.: Meddy trajectories in the Canary Basin measured during the SEMAPHORE experiment, 1993–1995, J. Geophys. Res. C, **103**, 25029 (1998).

133. Richardson, P.L., Bower, A.S., Zenk, W.: A census of Meddies tracked by floats, Prog. Oceanogr., **45**, 209 (2000).
134. Richardson, P.L., Hufford, G.E., Limeburner, R., Brown, W.S.: North Brazil current retroflection eddies, J. Geophys. Res. C, **99**, 5081 (1994).
135. Richardson, P.L., Mc Cartney, M.S., Maillard, C.: A search for Meddies in historical data, Dyn. Atmos. Oceans, **15**, 241 (1991).
136. Richardson, P.L., Walsh, D., Armi, L., Schröder, M., Price, J.F.: Tracking three meddies with SOFAR floats, J. Phys. Oceanogr., **19**, 371 (1989).
137. Ring Group (the): Gulf Stream cold-core rings: their physics, chemistry, and biology, Science, **212**, 4499, 1091 (1981).
138. Ripa, P.: On the stability of ocean vortices. In: Nihoul, J.C.J., Jamart, B.M. (eds.) Mesoscale/Synoptic Coherent Structures in Geophysical Turbulence, Vol. 50, Elsevier Oceanographic Series, Amsterdam, pp. 167–179 (1989).
139. Ripa, P.: General stability conditions for a multi-layer model, J. Fluid Mech., **222**, 119 (1991).
140. Robinson, A.R.: Eddies in Marine Science. 1st edn. Springer Verlag, Berlin, 609 pp. (1983).
141. Rossby, C.G.: On displacements and intensity changes of atmospheric vortices, J. Mar. Res., **7**, 175 (1948).
142. Saitoh, S., Kosaka, S., Iisaka, J.: Satellite infrared observations of Kuroshio warm-core rings and their application to study Pacific saury migration, Deep-Sea Res., **33**, 1601 (1986).
143. Serra, N.: Observation and Numerical Modeling of the Mediterranean Outflow. PhD thesis, University of Lisbon, 234pp. (2004).
144. Serra, N., Ambar, I.: Eddy generation in the Mediterranean undercurrent, Deep-Sea Res. II, **49** (19), 4225 (2002).
145. Schultz-Tokos, K., Hinrichsen, H.H., Zenk, W.: Merging and migration of two meddies, J. Phys. Oceanogr., **24**, 2129 (1994).
146. Shapiro, G.I., Meschanov, S.L., Yemel'Yanov, M.V.: Mediterranean lens after collision with seamounts, Oceanology, **32**, 279 (1992).
147. Spall, M.A., Robinson, A.R.: Regional primitive equation studies of the Gulf Stream meander and ring formation region, J. Phys. Oceanogr., **20**, 985 (1990).
148. Stegner, A., Zeitlin, V.: What can asymptotic expansions tell us about large-scale quasi-geostrophic anticyclonic vortices? Nonlinear Proc. Geophys., **2**, 186 (1995).
149. Stegner, A., Zeitlin, V.: Asymptotic expansions and monopolar solitary Rossby vortices in barotropic and two-layer models, Geophys. Astrophys. Fluid Dyn., **83**, 159 (1996).
150. Stern, M.E.: Entrainment of an eddy at the edge of a jet, J. Fluid Mech., **228**, 343 (1991).
151. Stern, M.E., Chassignet, E.P., Whitehead, J.A.: The wall jet in a rotating fluid, J. Fluid Mech., **335**, 1 (1997).
152. Sutyrin, G.G.: Critical effects of a seamount top on a drifting eddy, J. Mar. Res., **64**, 297 (2006).
153. Sutyrin, G.G., Hesthaven, J.S., Lynov, J.P., Rasmussen, J.J.: Dynamical properties of vortical structures on the beta-plane, J. Fluid Mech., **268**, 103 (1994).
154. Swaters, G.E.: On the baroclinic instability of cold-core coupled density fronts on a sloping continental shelf, J. Fluid Mech., **224**, 361 (1991).
155. Swaters, G.E.: On the baroclinic dynamics, Hamiltonian formulation. and general stability characteristics of density-driven surface currents. and fronts over a sloping continental shelf, Philos. Trans. Roy. Soc. Lond., **345A**, 295 (1993).
156. Swaters, G.E.: Numerical simulations of the baroclinic dynamics of density-driven coupled fronts and eddies on a sloping bottom, J. Geophys. Res., **103**, 2945 (1998).
157. Tang, B., Cushman-Roisin, B.: Two-layer geostrophic dynamics. Part II: Geostrophic turbulence, J. Phys. Oceanogr., **22**, 128 (1992).
158. Tychensky, A., Carton, X.J.: Hydrological and dynamical characterization of meddies in the Azores region: a paradigm for baroclinic vortex dynamics, J. Geophys. Res., **103**, 25, 061 (1998).

159. Vandermeirsch, F.O., Carton, X.J., Morel, Y.G.: Interaction between an eddy and a zonal jet. Part I. One and a half layer model, Dyn. Atmos. Oceans, **36**, 247 (2003).
160. Vandermeirsch, F.O., Carton, X.J., Morel, Y.G.: Interaction between an eddy and a zonal jet. Part II. Two and a half layer model, Dyn. Atmos. Oceans, **36**, 271 (2003).
161. van Geffen, J.H.G.M., Davies, P.A.: A monopolar vortex encounters an isolated topographic feature on a beta-plane, Dyn. Atmos. Oceans, **32**, 1 (2000).
162. van Kampen, N.G.: Elimination of fast variables, Phys. Rep., **124**, 69 (1985).
163. Verron, J.: Topographic eddies in temporally varying oceanic flows, Geophys. Astrophys. Fluid Dyn., **35**, 257 (1986).
164. Viudez, A., Dritschel, D.G.: Vertical velocity in mesoscale geophysical flows, J. Fluid Mech., **483**, 199 (2003).
165. Warn, T., Bokhove, O., Shepherd, T.G., Vallis, G.K.: Rossby-number expansions, slaving principles and balance dynamics, Q. J. Roy. Met. Soc., **121**, 723 (1995).
166. Watts, D.R., Bane, J.M., Tracey, K.L., Shay, T.J.: Gulf Stream path and thermocline structure near 74 °W and 68 °W, J. Geophys. Res. C, **100**, 18291 (1995).
167. Yasuda, I.: Geostrophic vortex merger and streamer development in the ocean with special reference to the Merger of Kuroshio warm core rings, J. Phys. Oceanogr., **25**, 979 (1995).
168. Yasuda, I., Okuda, K., Hirai, M.: Evolution of a Kuroshio warm-core ring – variability of the hydrographic structure, Deep-Sea Res., **39**, S131 (1992).
169. Yavneh, I., Mc Williams, J.C.: Breakdown of the slow manifold in the shallow-water equations, Geophys. Astrophys. Fluid Dyn., **75**, 131 (1994).
170. Yavneh, I., Mc Williams, J.C.: Robust multigrid solution of the shallow-water balance equations, J. Comp. Phys., **119**, 1 (1995).
171. Yavneh, I., Shchepetkin, A., Mc Williams, J.C., Graves, L.P.: Multigrid solution of rotating stably-stratified flows, J. Comp. Phys., **136**, 245 (1997).
172. Zhurbas, V.M., Lozovatskiy, I.D., Ozmidov, R.V.: The effect of seamounts on the propagation of meddies in the Atlantic Ocean, Doklady Akad. Nauk SSSR, **318**, 1224 (1991).
173. Zubin, A.B., Ozmidov, R.V.: A lens of Mediterranean water in the vicinity of the ampere and Josephine seamounts, Doklady Akad. Nauk SSSR, **292**, 716 (1987).

Chapter 4
Lagrangian Dynamics of Fronts, Vortices and Waves: Understanding the (Semi-)geostrophic Adjustment

V. Zeitlin

The geostrophic adjustment, i.e. the relaxation of the rotating stratified fluid to the geostrophic equilibrium is a key process in geophysical fluid dynamics. We study it in idealized plan-parallel and axisymmetric configurations (semi-geostrophic adjustment) in a hierarchy of models of increasing complexity: rotating shallow water equations, two-layer rotating shallow water equations, and continuously stratified hydrostatic Boussinesq equations. We show that the use of Lagrangian variables allows for substantial advances in understanding the semigeostrophic adjustment and related issues: existence of the adjusted state ("slow manifold"), wave emission, wave trapping, and wave breaking, pulsating front solutions, symmetric/inertial instability, and frontogenesis.

4.1 Introduction: Geostrophic Adjustment in GFD and Related Problems

Geostrophic adjustment, i.e. relaxation of the rotating fluid to the state of geostrophic equilibrium (equilibrium between the pressure and the Coriolis forces) is a key process in geophysical fluid dynamics (GFD), cf, e.g. Blumen [3]. The so-called balanced states, close to the equilibrium and associated with frontal and vortex structures in the atmosphere and oceans, evolve slowly, in contradistinction with fast unbalanced motions associated with waves. The dynamical separation ("splitting") of balanced and unbalanced motions in GFD is of utmost importance for applications, such as weather and climate predictions. A concise introduction to the dynamical splitting of fast and slow motions with references may be found in Reznik and Zeitlin [19].

In rotating stratified fluids the geostrophic balance (the "geostrophic wind" relation) is to be combined with the hydrostatic balance giving the so-called thermal wind relation. The process of relaxation to the balanced state is still called the

V. Zeitlin (✉)

LMD, Ecole Normale Supérieure, 24 rue Lhomond, 75005 Paris Cedex 05, France,
zeitlin@lmd.ens.fr

Zeitlin, V.: *Lagrangian Dynamics of Fronts, Vortices and Waves: Understanding the (Semi-)geostrophic Adjustment*. Lect. Notes Phys. **805**, 109–137 (2010)
DOI 10.1007/978-3-642-11587-5_4   © Springer-Verlag Berlin Heidelberg 2010

geostrophic adjustment. We should note in passing that the thermal wind relation alone allows to understand many of the observed synoptic-scale features in the atmosphere and oceans [16, 11].

In the fluid dynamics perspective, a series of questions arise in what concerns the process of adjustment. The first is whether the adjusted state exists. If not, what will be the end state of the evolution and may the adjustment process lead to a singularity? If the adjusted state does exist, is it attainable, or in other words, is the adjustment complete? What happens if the adjusted state is unstable? The details of the adjustment process are also of importance: how the energy is evacuated via the unbalanced wave motions? What are the properties of the emitted waves?

In what follows we will show that the Lagrangian approach to idealized configurations of straight fronts and circular vortices allows to substantially advance in understanding the process of adjustment and, in many cases, to give exhaustive answers to the above-posed questions. The major simplification arises from the independence of the system of one of the spatial coordinates. In this case the adjusted states are not just slow, but stationary ("infinitely slow"), and the introduction of Lagrangian coordinates considerably simplifies the problem.

This chapter is organized as follows. We start in Sect. 4.2 from the simplest, albeit conceptually most important model of GFD: the rotating shallow water model (RSW) and show how the adjustment problem may be solved in its 1.5-dimensional version using Lagrangian coordinates. We then introduce in Sect. 4.3 a rudimentary stratification by superimposing two shallow water layers and display the novel phenomena arising in this case. Finally, in Sect. 4.4 we analyse the continuously stratified, so-called primitive equations of 2.5-dimensional GFD. In all of the above-mentioned models the "half-" dimensionality means that although the dependence of all dynamical variables of one of the spatial coordinates is removed, the non-zero velocity in this passive direction is still allowed. The presentation in Sect. 4.2 is based on Zeitlin et al. [25], that of Sect. 4.3 on LeSommer et al. [14] and on Zeitlin [24], and that of Sect. 4.4 on Plougonven and Zeitlin [17], although new with respect to the above-mentioned papers added in each section.

4.2 Fronts, Waves, Vortices and the Adjustment Problem in 1.5d Rotating Shallow Water Model

4.2.1 The Plane-Parallel Case

4.2.1.1 General Features of the Model

The RSW equations in the f-plane approximation with no dependence on the y-coordinates (i.e. $\partial_y \ldots \equiv 0$) are

$$
\begin{aligned}
\partial_t u + u \partial_x u - f v + g \partial_x h &= 0, \\
\partial_t v + u \partial_x v + f u &= 0, \\
\partial_t h + \partial_x (u h) &= 0 .
\end{aligned}
\tag{4.1}
$$

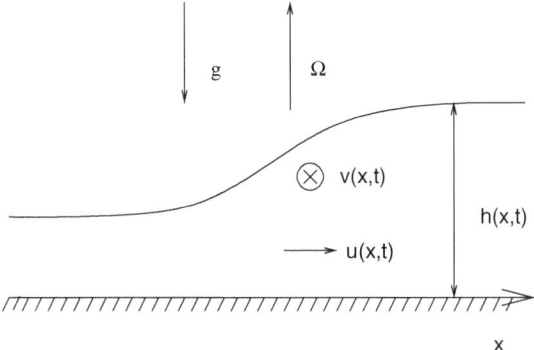

Fig. 4.1 Schematic representation of the 1.5d RSW model

Here u, v are the across-front and the along-front components of the velocity, respectively, h is the total depth (no topographic effects will be considered in what follows), g is gravity (or reduced gravity – see below), f is the Coriolis parameter, which will be supposed constant (the f-plane approximation), unless the opposite is explicitly stated, and the subscripts denote the corresponding partial derivatives. A sketch of the plane-parallel RSW configuration is presented in Fig. 4.1.

The model possesses two Lagrangian invariants: the generalized (geostrophic) momentum $M = v + fx$ and the potential vorticity (PV) $Q = \frac{v_x + f}{h}$:

$$(\partial_t + u\partial_x)M = 0, \quad (\partial_t + u\partial_x)Q = 0, \tag{4.2}$$

which are related: $Q = \frac{\partial_x M}{h}$. Let us emphasize that the conservation of the geostrophic momentum is a consequence of 1.5 dimensionality of the problem. The straightforward linearization around the state of rest $h = H_0 = constant$ gives the zero-frequency (*slow*) mode (the linearized PV) and the *fast* surface inertia - gravity waves with the dispersion law:

$$\omega = \pm(c_0^2 k^2 + f^2)^{\frac{1}{2}}, \tag{4.3}$$

where $c_0 = \sqrt{gH_0}$ is the "sound speed", i.e. the maximum phase speed of short inertia-gravity waves, ω is the frequency and k is the wavenumber.

The *geostrophic equilibria* are steady states:

$$fv = g\partial_x h. \tag{4.4}$$

They are the exact solutions of the full nonlinear equations (4.1), which makes a difference with respect to the full 2d RSW equations, where the geostrophic equilibria are not solutions, but are just slow (e.g. Reznik et al. [20]).

4.2.1.2 Lagrangian Approach to 1.5d RSW

In order to fully exploit the existence of a pair of Lagrangian invariants in the model, it is natural to introduce the Lagrangian coordinates $X(x, t)$ of the fluid "parcels" (in fact, fluid lines along the y-axis). They are given by the mapping $x \to X(x, t)$, where x is a fluid parcel position at $t = 0$ and X – its position at time t. Hence $\dot{X} \equiv \partial_t X = u(X, t)$. The momentum equations in (4.1) become:

$$\ddot{X} - fv + g\frac{\partial h}{\partial X} = 0, \tag{4.5}$$

$$\partial_t (v + fX) = 0, \tag{4.6}$$

where v is considered as a function of x and t. The mass conservation for each fluid element $h(X, t)dX = h_I(x)dx$ means that

$$h(X, t) = h_I(x)\frac{\partial x}{\partial X}. \tag{4.7}$$

This equation, obviously, is equivalent to the continuity equation in (4.1). Equation (4.6) immediately gives

$$v(x, t) + fX(x, t) = v_I(x)fx = M(x). \tag{4.8}$$

By applying the chain differentiation rule to (4.7) and injecting the result into (4.5) we get a closed equation for X:

$$\ddot{X} + f^2 X + gh'_I\frac{1}{(X')^2} + \frac{gh_I}{2}\left[\frac{1}{(X')^2}\right]' = fM, \tag{4.9}$$

where prime denotes ∂_x. In terms of the deviations of fluid parcels from their initial positions $X(x, t) = x + \phi(x, t)$ (4.9) takes the form:

$$\ddot{\phi} + f^2\phi + gh'_I\frac{1}{(1 + \phi')^2} + \frac{1}{2}gh_I\left[\frac{1}{(1 + \phi')^2}\right]' = fv_I. \tag{4.10}$$

This single equation is equivalent to the whole system (4.1). It should be solved with initial conditions $\phi(t = 0) = 0$; $\dot{\phi}(t = 0) = u_I(x)$. Thus, the Cauchy (adjustment) problem is well and naturally posed for this equation.

It should be noted that 1.5d RSW in Lagrangian variables may be as well formulated in the β-plane approximation, i.e. taking into account the dependence of the Coriolis parameter on latitude: $f = f_0 + \beta y$. For example, for purely zonal flows on the equatorial β-plane ($f_0 \equiv 0$) we get

$$\ddot{Y} + \beta Y u + g \frac{\partial h}{\partial Y} = 0,$$ (4.11)

$$\partial_t \left(u - \beta \frac{Y^2}{2} \right) = 0,$$

$$h(Y,t) = h_I(y)\frac{\partial y}{\partial Y},$$ (4.12)

and the closed equation for Y follows:

$$\ddot{Y} + \beta Y \left(u_I + \beta \frac{Y^2 - y^2}{2} \right) + gh'_I \frac{1}{(Y')^2} + \frac{gh_I}{2}\left[\frac{1}{(Y')^2} \right]' = 0,$$ (4.13)

to be solved with initial conditions $Y(y,0) = y$, $\dot{Y}(y,0) = v_I(y)$.

4.2.1.3 The Slow Manifold

By additional change of variables $x = x(a)$, the elevation profile in (4.5), (4.6), and (4.7) may be "straightened" to a uniform height H in order to have $J = \frac{\partial X}{\partial a} = \frac{H}{h(X,t)}$. It is easy to see that $\frac{\partial h}{\partial X} = \frac{\partial P}{\partial a}$, where $P = \frac{gH}{2J^2}$ is the so-called Lagrangian pressure variable. The Lagrangian equations of motion then take the form:

$$\dot{u} - fv + gH\frac{\partial}{\partial a}\frac{1}{2J^2} = 0,$$ (4.14)

$$\dot{v} + fu = 0,$$ (4.15)

$$\dot{J} - \partial_a u = 0,$$ (4.16)

and may be again reduced to a single equation:

$$\ddot{J} + f^2 J + \frac{\partial^2 P}{\partial a^2} = fHQ,$$ (4.17)

where Q – potential vorticity as a function of the a variable is $Q(a)$ $= \frac{1}{H}\left(\frac{\partial v}{\partial a} + fJ \right) = \frac{1}{H}\left(\frac{\partial v_I}{\partial a} + fJ_I \right)$.

The slow manifold is the stationary solution of (4.17) or (4.9). By re-introducing the X-variable and the dependent variable $\eta = \frac{h}{H}$ we get

$$-\frac{g}{f}\frac{d^2 h(X)}{dX^2} + h(X)Q(X) = -f.$$ (4.18)

Note that potential vorticity in terms of initial height and velocity fields reads $Q(X(x)) = \frac{f + \frac{\partial v_I}{\partial x}}{h_I}$. The following theorem may be proved by standard methods of ordinary differential equations (Zeitlin et al. [25]): *Equation (4.18) has a bounded and everywhere positive unique solution $h(X)$ on $\mathbf{R}$ for positive $Q(X)$ with compact support and constant asymptotics (frontal case).*

It should be noted that positiveness of Q corresponds to the absence of the so-called inertial instability (see the next section). The latter is related to the presence of sub-inertial (i.e. $\omega < f$) frequencies in the spectrum of small excitations of the adjusted state. It may be, however, explicitly shown either in Eulerian variables (Zeitlin et al. [25]) or in Lagrangian variables (see below) that the spectrum of small perturbations over an adjusted front in 1.5d RSW is supra-inertial. Although we have no proof for non-positive distributions of Q, direct numerical simulations (Bouchut et al. [4]) indicate that a unique adjusted state is always achieved in this case too.

4.2.1.4 Relaxation Towards the Adjusted State

Once the existence of the adjusted state is established, the process of relaxation towards this state may be analysed. The first step in studying relaxation is linearization around the adjusted state:

$$u = \tilde{u}, \quad v = v_s + \tilde{v}, \quad J = J_s + \tilde{J}$$

$$\partial_t \tilde{u} - f\tilde{v} - gH\partial_a(\tilde{J}/J_s^3) = 0, \tag{4.19}$$

$$\partial_t \tilde{v} + fu = 0, \tag{4.20}$$

$$\partial_t \tilde{J} - \partial_a u = 0, \tag{4.21}$$

where the Lagrangian time derivative is denoted by ∂_t from now on. By using

$$f\tilde{J} + \partial_a \tilde{v} = 0, \tag{4.22}$$

it is easy to get a single equation for $\tilde{J}$ and/or for $\tilde{v}$

$$\partial_{tt}^2 \tilde{J} + f^2 \tilde{J} - gH\partial_{aa}^2(\tilde{J}/J_s^3) = 0, \quad \partial_{tt}^2 \tilde{v} + f^2 \tilde{v} - gH\partial_a(\tilde{v}_a/J_s^3) = 0. \tag{4.23}$$

Let us consider stationary solutions

$$\tilde{J} = \hat{J}(a)e^{-i\omega t} + \text{c.c.}, \quad \tilde{v} = \hat{v}(a)e^{-i\omega t} + \text{c.c.}. \tag{4.24}$$

Then the stationary equations are

$$\partial_{aa}^2(gH_s\hat{J}) + (\omega^2 - f^2)\hat{J} = 0, \tag{4.25}$$

$$\partial_a(gH_s\partial_a\hat{v}) + (\omega^2 - f^2)\hat{v} = 0, \tag{4.26}$$

where we denoted $H_s = H/J_s^3$. The equation for $\hat{v}$ is self-adjoint and *supra-inertiality* of ω and, hence, *the absence of trapped states* follows trivially from (4.26) by multiplying by $\hat{v}^*$ and integrating by parts:

$$\omega^2 = f^2 + \frac{\int da\, g H_s \left|\partial_a \hat{v}\right|^2}{\int da\, |\hat{v}|^2} \quad \Rightarrow \omega^2 \geq f^2. \tag{4.27}$$

By using a new dependent variable

$$\hat{v} = \frac{\psi}{g H_s^{1/2}}, \tag{4.28}$$

we transform the stationary equation to a two-term canonical form

$$\frac{d^2\psi}{da^2} + \left[\frac{\omega^2 - f^2}{g H_s} - \frac{1}{4} \left(\frac{(H_s)_a}{H_s} \right)^2 - \frac{1}{2} \left(\frac{(H_s)_a}{H_s} \right)_a \right] \psi = 0. \tag{4.29}$$

Rewritten as

$$\frac{d^2\psi}{da^2} + k_\psi^2(a)\psi = 0, \tag{4.30}$$

this equation can be interpreted as that of a quantum mechanical oscillator with variable frequency $k_\psi(a)$ (or as a Schrödinger equation with a potential V and an energy E such that $k_\psi^2 = E - V(a)$). It is clear that k_ψ^2 can be negative for $\omega > f$ and suitable H_s. This means that for certain intervals on the x-axis the wavenumber k_ψ may be imaginary and, hence, quasi-stationary states slowly tunneling out such zones may exist. Thus, the wave motions can be maintained for long times in such locations.

4.2.1.5 Wave Breaking

The direct simulations of the Lagrangian equations of motion indicate that singularities (shocks) may appear in the emitted inertia-gravity field. In the context of adjustment, shocks could provide an alternative sink of energy, whence the importance to establish the criteria of wave breaking and shock formation. Shocks are of no surprise in gas dynamics, and the shallow-water equations are a particular case of it. The only question, thus, is the role of rotation in this process. The Lagrangian approach, again, proves to be efficient (Zeitlin et al. [25]). The dimensionless Lagrangian equations of motion in a-variables introduced above are

$$\partial_t u + \partial_a p = v\,,$$
$$\partial_t J - \partial_a u = 0\,, \tag{4.31}$$

where v is not an independent variable and is to be found from $\partial_a v = Q(a) - J$. We thus have a quasi-linear system

$$\partial_t \begin{pmatrix} u \\ J \end{pmatrix} + \begin{pmatrix} 0 & -J^{-3} \\ -1 & 0 \end{pmatrix} \partial_a \begin{pmatrix} u \\ J \end{pmatrix} = \begin{pmatrix} v \\ 0 \end{pmatrix}. \tag{4.32}$$

The eigenvalues of the matrix in the l.h.s. of (4.32) are $\mu_\pm = \pm J^{-\frac{3}{2}}$ and the corresponding left eigenvectors are $\left(1, \pm J^{-\frac{3}{2}}\right)$. Hence, Riemann invariants are $w_\pm = u \pm 2J^{-\frac{1}{2}}$ and we have

$$\partial_t w_\pm + \mu_\pm \partial_a w_\pm = v. \tag{4.33}$$

Expressions of original variables in terms of $w_\pm$ are

$$u = \frac{1}{2}(w_+ + w_-), \ J = \frac{16}{(w_+ - w_-)^2} > 0, \ \mu_\pm = \pm \left(\frac{w_+ - w_-}{4}\right)^3. \tag{4.34}$$

In terms of the derivatives of the Riemann invariants $r_\pm = \partial_a w_\pm$, we get

$$\partial_t r_\pm + \mu_\pm \partial_a r_\pm + \frac{\partial \mu_\pm}{\partial w_+} r_+ r_\pm + \frac{\partial \mu_\pm}{\partial w_-} r_- r_\pm = \partial_a v = Q(a) - J, \tag{4.35}$$

which may be rewritten using Lagrangian derivatives along the characteristics $\frac{d}{dt_\pm} = \partial_t + \mu_\pm \partial_a$ as

$$\frac{dr_\pm}{dt_\pm} + \frac{\partial \mu_\pm}{\partial w_+} r_+ r_\pm + \frac{\partial \mu_\pm}{\partial w_-} r_- r_\pm = Q(a) - J. \tag{4.36}$$

Wave breaking and shock formation correspond to $r_\pm \to \pm\infty$ in finite time.

In terms of new variables $R_\pm = e^\lambda r_\pm$, with $\lambda = \frac{3}{128} \log |w_+ - w_-|$, (4.35) may be rewritten as

$$\frac{dR_\pm}{dt_\pm} = -e^{-\lambda} \frac{\partial \mu_\pm}{\partial w_\pm} R_\pm^2 + e^\lambda \left(Q(a) - J\right), \tag{4.37}$$

where $\frac{\partial \mu_\pm}{\partial w_\pm} = \frac{3}{64}(w_+ - w_-)^2 > 0$.

The qualitative analysis of these generalized Ricatti equations shows that if initial relative vorticity $Q - J = \partial_a v$ is sufficiently negative (anti-cyclonic), rotation does not stop wave breaking, which is taking place for any initial conditions. However, if the relative vorticity is positive (cyclonic case), as well as the derivatives of the Riemann invariants at the initial moment, there is no breaking. An example of wave breaking due to the geostrophic adjustment of the unbalanced jet is presented in Fig. 4.2.

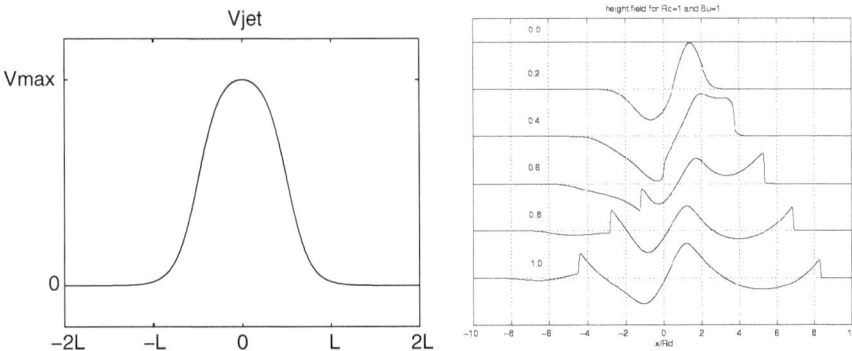

Fig. 4.2 Wave breaking and shock formation (*right panel*) during adjustment of the unbalanced jet (*left panel, top to bottom*: consecutive profiles of the free surface with time measured in f^{-1} units). Length is measured in deformation radius units: $L = R_d = \frac{gH}{f}$

4.2.1.6 "Trapped Waves" in 1.5d RSW: Pulsating Density Fronts

The above-established supra-inertiality of the spectrum of the small perturbations around a balanced 1.5d RSW front means the absence of trapped waves, and, hence, the attainability of the adjusted state by evacuating the excess of energy via inertia-gravity wave emission (eventually with shock formation). There exist, however, the RSW fronts, where the wave emission is impossible. These are the lens-type configurations with terminating profile of fluid height. Such RSW configurations are used to model oceanic double density fronts, either outcropping or incropping, e.g. Griffiths et al. [10]. In Lagrangian description (4.9) the evolution of a double RSW front corresponds to positive h_I terminating at $x = x_\pm$. Adjustment of such fronts, therefore, should proceed without outward IGW emission. An example of adjusted front treated in literature is given in Fig. 4.3.

A family of exact unbalanced pulsating solutions is known for such fronts (Frei [9]; Rubino et al. [22]). Let us make the following ansatz:

$$X(x,t) = x\chi(t), \quad h_I(x) = \frac{h_0}{2}\left(1 - \frac{x^2}{L^2}\right), \quad v_I(x) = x\Omega, \qquad (4.38)$$

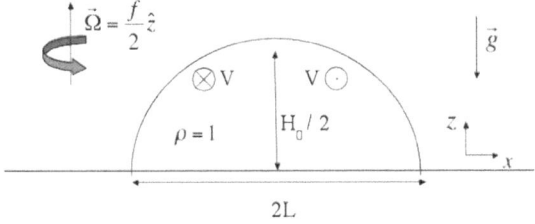

Fig. 4.3 An example of equilibrated double density front

where h_0, Ω, L are constants. Plugging (4.38) into (4.9) and non-dimensionalizing with the timescale f^{-1} and the length-scale L gives the following ODE for χ:

$$\ddot{\chi} + \chi - \frac{\gamma}{\chi^2} = \mu, \tag{4.39}$$

where γ is the Burger number $\frac{gh_0}{f^2 L^2}$ and $\mu = 1 + \frac{\Omega}{f}$.

Integrating (4.39) once gives

$$\frac{\dot{\chi}^2}{2} + P(\chi) = E, \quad P(\chi) = \frac{\chi^2}{2} - \mu\chi + \frac{\gamma}{\chi}, \tag{4.40}$$

where the integration constant E is expressed in terms of initial conditions $\chi(t = 0) = 1$, $\dot{\chi}(t = 0) = U$:

$$E = \frac{U^2}{2} + \frac{1}{2} - \mu + \gamma. \tag{4.41}$$

Equation (4.40) may be integrated in elliptic functions. The "potential" $P(\chi)$ being convex, the solution for χ is finite amplitude and oscillating with *supra-inertial frequency*. The minimum of P corresponds to the front in geostrophic equilibrium and constant $\chi = 1$. Thus, the adjustment (initial-value) problem for double density fronts will result, in general, in a pulsating solution, whereas relaxation to the steady state is possible only due to viscous effects (shocks).

4.2.2 Axisymmetric Case

4.2.2.1 Governing Equations and Lagrangian Invariants

Axisymmetric RSW motion is described in cylindrical coordinates by fields depending on radial variable only. As in the rectilinear case, it is possible to reduce the whole dynamics to a single PDE for a Lagrangian variable $R(r, t)$, the distance to the center of a "particle" (or rather a particle ring) initially situated at r.

We first rewrite the Eulerian RSW equations in cylindrical coordinates (r, θ) and assume exact axial symmetry:

$$(\partial_t + u_r \partial_r)u_r - u_\theta \left(f + \frac{u_\theta}{r} \right) + \partial_r h = 0 ,$$

$$(\partial_t + u_r \partial_r)u_\theta + u_r \left(f + \frac{u_\theta}{r} \right) = 0 , \tag{4.42}$$

$$\partial_t h + \frac{1}{r} \partial_r (r \, u_r \, h) = 0 .$$

Here u_r, u_θ are the radial and azimuthal components of velocity. Note that the adjusted stationary state changes character as compared to the rectilinear case: it verifies conditions of the *cyclo-geostrophic balance* and not of the purely geostrophic one:

$$u_\theta \left(f + \frac{u_\theta}{r} \right) = \partial_r h, \; u_r = 0. \tag{4.43}$$

Multiplying the second equation in (4.42) by r, we recover the conservation of angular momentum:

$$(\partial_t + u_r \, \partial_r) \left(r u_\theta + f \frac{r^2}{2} \right) = 0 , \tag{4.44}$$

which replaces the conservation of geostrophic momentum in the plane-parallel case. Equation (4.42) can be rewritten using the Lagrangian coordinate $R(r, t)$. Integrating (4.44) gives

$$R(r, t) \, u_\theta(r, t) + f \frac{R^2(r, t)}{2} = r \, u_{\theta I}(r) + f \frac{r^2}{2} \equiv G(r) , \tag{4.45}$$

where $u_{\theta I}$ is the initial azimuthal velocity profile. Using the above expression we get

$$u_\theta \left(f + \frac{u_\theta}{R} \right) = \frac{1}{R} \left(G - f \frac{R^2}{2} \right) \left(f + \frac{G}{R^2} - \frac{f}{2} \right)$$

$$= \frac{1}{R^3} \left(G^2 - \frac{f^2 R^4}{4} \right) . \tag{4.46}$$

The mass conservation is expressed by the following relation:

$$h(r, t) \, R(r, t) \, dR = h_I(r) r \, dr . \tag{4.47}$$

With the help of (4.46), (4.47) and the definition $\dot{R}(r, t) = u_r(r, t)$, the radial momentum equation becomes

$$\ddot{R} + \frac{f^2}{4} R - \frac{1}{R^3} G^2 + \frac{1}{\partial_r R} \partial_r \left(\frac{r \, h_I}{R \, \partial_r R} \right) = 0 , \tag{4.48}$$

to be solved with initial conditions $R(r, 0) = r$, $\dot{R}(r, 0) = u_{r_I}$. The stationary part of this equation defines the adjusted, slow states. The fast motions are axisymmetric IGW. Indeed, for small perturbations about the state of rest:

$$R(r, t) = r + \phi(r, t) , \tag{4.49}$$

with $|\phi| \ll r$, $h_I(r) = 1$ and $u_{\theta I}(r) = 0$, the following equation is obtained after some algebra:

$$\ddot{\phi} + f^2\,\phi - \frac{\partial_r\phi}{r} - \partial_{rr}^2\phi + \frac{\phi}{r^2} = 0\,. \tag{4.50}$$

If solutions are sought in the form $\phi(r,t) = \hat{\phi}(r)\,e^{i\omega t}$, (4.50) yields, after a change of variables, the canonical equation for the Bessel functions. The familiar axisymmetric wave solutions involving Bessel functions J_1 then follow:

$$\phi(r,t) = C\,J_1(\sqrt{\omega^2 - f^2}\,r)\,e^{i\omega t} + c.c.\,, \tag{4.51}$$

where C is the wave amplitude.

The whole program of the previous section may be carried on as well in cylindrical coordinates, with similar conclusions. We present below only the case of the axisymmetric density fronts (Sutyrin and Zeitlin [23]).

4.2.2.2 Axisymmetric Density Fronts and Radial "pulson" solutions

We make the following ansatz in (4.48):

$$h_I(r) = \frac{h_0}{2}\left(1 - \frac{r^2}{L^2}\right),\quad R(r,t) = r\phi(t),\quad u_{\theta I}(r) = r\Omega,\ \Omega = \text{const.} \tag{4.52}$$

Then by non-dimensionalizing the system in the same way as for the rectilinear fronts, introducing the Burger number γ, and denoting $M = \frac{1}{2} + \frac{\Omega}{f}$ we get

$$\ddot{\phi} + \frac{\phi}{4} - \frac{M^2}{\phi^3} - \frac{\gamma}{\phi^3} = 0, \tag{4.53}$$

to be solved with initial conditions $\phi(0) = 1$, $\dot{\phi}(0) = u_{r_I}$. A drastic simplification of this equation is provided by the substitution $\phi^2 = \chi$ which immediately gives the equation of the harmonic oscillator with shifted equilibrium position:

$$\ddot{\chi} + \chi - 4E = 0,\quad E = \frac{u_{r_I}^2}{2} + \frac{1}{8} + \frac{M^2 + \gamma}{2} > 0. \tag{4.54}$$

The "radial pulson" solution (cf. Rubino et al. [21] for a derivation in Eulerian framework) satisfies the initial conditions $\chi(0) = 1$, $\dot{\chi}(0) = 2u_{r_I}$ and is given by

$$\chi(t) = 4E + (1 - 4E)\cos t + (2u_{r_I} + 1 - 4E)\sin t. \tag{4.55}$$

The crucial difference between the radial and rectilinear pulson, thus, is that the former always has inertial frequency and thus represents nonlinear inertial oscillations, while the latter is always supra-inertial.

4.3 Including Baroclinicity: 2-Layer 1.5d RSW

4.3.1 Plane-Parallel Case

4.3.1.1 Governing Equations and General Properties of the Model

To introduce the baroclinic effects in the dynamics in the simplest way we consider the two-layer rotating shallow water model. We use the rigid lid upper boundary condition and again consider for simplicity a flat bottom. In this case the equations governing the motion of two superimposed rotating shallow-water layers of unperturbed depths $H_{1,2}$, $H_1 + H_2 = H$ and densities $\rho_{1,2}$ in Cartesian coordinates under hypothesis of no dependence of y (straight two-layer fronts) are

$$\partial_t u_i + u_i \partial_x u_i - f v_1 + \rho_i^{-1} \partial_x \pi_i = 0, \tag{4.56a}$$

$$\partial_t v_i + u_i (f + \partial_x v_i) = 0, \tag{4.56b}$$

$$\partial_t h_i + \partial_x ((h_i u_i) = 0, \quad i = 1, 2 \tag{4.56c}$$

$$\pi_1 + g'(\rho_1 h_1 + \rho_2 h_2) = \pi_2, \tag{4.56d}$$

$$h_1 + h_2 = 1, \tag{4.56e}$$

where no sum over repeated index is understood, π_i are the pressures in the layers, $g' = \frac{\rho_2 - \rho_1}{\rho_2 + \rho_1} g$ is the reduced gravity and h_i are the variable layers depths. A sketch of the 2-layer 1.5d RSW is presented in Fig. 4.4.

The Lagrangian invariants of equations (4.56a), (4.56b) and (4.56c) are potential vorticities and geostrophic momenta in each layer:

$$Q_i = \frac{f + \partial_x v_i}{h_i}, \quad M_i = f x + \partial_x v_i, \quad i = 1, 2. \tag{4.57}$$

For any solution of system (4.56a), (4.56b), (4.56c), (4.56d) and (4.56e), constraint (4.56e) imposes that

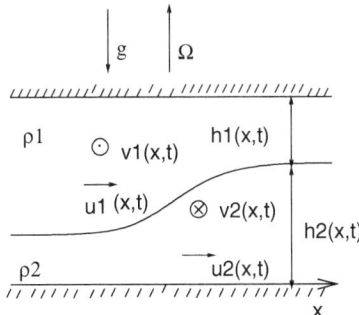

Fig. 4.4 Schematic representation of the 2-layer 1.5d RSW model

$$\partial_x (h_1 u_1 + h_2 u_2) = 0. \tag{4.58}$$

Hence, the barotropic across-front velocity is

$$U = \frac{h_1 u_1 + h_2 u_2}{H} = U(t). \tag{4.59}$$

Choosing the boundary condition of absence of the mass flux across the front sets $U = 0$. The geostrophic equilibria are stationary solutions:

$$u_i = 0, \; v_i = \frac{1}{f \, \rho_i} \, \partial_x \pi_i \, , i = 1, 2 \, , \pi_2 = \pi_1 + g(\rho_1 h_1 + \rho_2 h_2). \tag{4.60}$$

The fast motions in the linear approximation are internal inertia-gravity waves propagating along the interface between the layers. By linearizing about the rest state $h_1 = H_1$, $h_2 = H_2$, $u_{1,2} = 0$, $v_{1,2} = 0$, the dispersion relation for the waves with frequency ω and wavenumber k follows:

$$\omega^2 (\omega^2 - f^2 - c_e^2 k^2) = 0. \tag{4.61}$$

Here $c_e^2 = g' H_e$ is the phase speed of the waves, $H_e = \frac{(\rho_2 - \rho_1) H_1 H_2}{\rho_1 H_1 + \rho_2 H_2}$ is the equivalent height for the baroclinic modes of the model. As in the one-layer model, conditions for existence and uniqueness of the adjusted state can be obtained as conditions for existence and uniqueness of solutions to the PV equations (LeSommer et al. [14]). These equations can be combined to give two ordinary differential equations for the depths of the layers:

$$\frac{g'}{f} h_1'' - (Q_2 + r \, Q_1) h_1 = -(-f(1-r) + H \, Q_2) \, , \tag{4.62a}$$

$$\frac{g'}{f} h_2'' - (Q_2 + r \, Q_1) h_2 = -(f(1-r) + r H \, Q_1) \, , \tag{4.62b}$$

where notation $r = \rho_1 / \rho_2$ for the density ratio of the layers has been introduced and the prime denotes the x'- differentiation. An essential difference of these equations from their one-layer counterpart is that the forcing terms at the r.h.s. are not constant. They, nevertheless, may be analysed by the same method as in 1dRSW.

For an equation of the form $h'' - R(x) \, h = -S(x)$, the existence and uniqueness of solutions are guaranteed if R and S have constant asymptotics at $\pm\infty$. Furthermore, the solution is positive if R and S are positive. Hence, for the initial states with localized PV anomalies such that

$$Q_1 \geq 0 \quad \text{and} \quad Q_2 \geq (1-r) \, f/H \, , \tag{4.63}$$

the above equations have unique solutions h_1 and h_2 that are everywhere positive.

A crucial simplification of the rigid-lid 2-layer equations follows from the fact the pressures π_i may be eliminated from (4.56a), (4.56b) and (4.56c). Indeed by using (4.58) and (4.56e) and (4.56d) we get, again under the hypothesis of zero overall across-front mass flux:

$$
\frac{\partial \pi_1}{\partial x} = \left(\frac{h_1}{\rho_1} + \frac{h_2}{\rho_2}\right)^{-1} \left(f(h_1 v_1 + h_2 v_2) - \frac{\partial}{\partial x}\left(h_1 u_1^2 + h_2 u_2^2\right) \right.
$$
$$
\left. - \frac{g h_2}{\rho_2}\frac{\partial}{\partial x}(\rho_1 h_1 + \rho_2 h_2)\right), \tag{4.64}
$$

$$
\frac{\partial \pi_2}{\partial x} = \left(\frac{h_1}{\rho_1} + \frac{h_2}{\rho_2}\right)^{-1} \left(f(h_1 v_1 + h_2 v_2) - \frac{\partial}{\partial x}\left(h_1 u_1^2 + h_2 u_2^2\right) \right.
$$
$$
\left. + \frac{g h_1}{\rho_1}\frac{\partial}{\partial x}(\rho_1 h_1 + \rho_2 h_2)\right). \tag{4.65}
$$

One can use (4.64), (4.65) in order to reduce the system to four equations for four independent variables u_2, h_2, v_2 and v_1, i.e. lower (heavier)-layer variables plus upper-layer jet velocity:

$$
\frac{\partial u_2}{\partial t} + u_2\frac{\partial u_2}{\partial x} - f v_2 + \frac{\rho_1}{\rho_2 h_1 + \rho_1 h_2}\left(f(h_1 v_1 + h_2 v_2)\right.
$$
$$
\left. - \frac{\partial}{\partial x}\left(h_1 u_1^2 + h_2 u_2^2\right) + \frac{g(\rho_2 - \rho_1)}{\rho_1}h_1\frac{\partial h_2}{\partial x}\right) = 0, \tag{4.66}
$$

$$
\frac{\partial h_2}{\partial t} + u_2\frac{\partial h_2}{\partial x} + h_2\frac{\partial u_2}{\partial x} = 0, \tag{4.67}
$$

$$
\frac{\partial v_2}{\partial t} + u_2\frac{\partial v_2}{\partial x} + f u_2 = 0, \tag{4.68}
$$

$$
\frac{\partial v_1}{\partial t} + u_2\frac{\partial v_1}{\partial x} + (u_1 - u_2)\frac{\partial v_1}{\partial x} + f u_1 = 0, \tag{4.69}
$$

where

$$
u_1 = \frac{h_2 u_2}{h_2 - H}, \quad h_1 = H - h_2. \tag{4.70}
$$

4.3.1.2 Lagrangian Approach to 2-Layer 1.5d RSW

We start from the system (4.66), (4.67), (4.68), (4.69) and (4.70), taken for simplicity in the frequently used limit $r \to 1$ and introduce the Lagrangian coordinate

$X(x, t)$ corresponding to the positions of the fluid particles in the lower layer. In terms of displacements ϕ with respect to initial positions $X(x, t) = x + \phi(x, t)$. The corresponding Lagrangian derivative is $\frac{d}{dt} = \frac{\partial}{\partial t} + u_2 \frac{\partial}{\partial x}$. The dependence of the height variable h_2 on the Lagrangian labels and transformation of its derivatives are obtained via the mass conservation in the lower layer: $h_{2_I} dx = h_2(X(x,t), t) dX$. The subscript 2 will be omitted in what follows. As in the one-layer case, (4.68) expresses the conservation of the geostrophic momentum in the lower layer and allows to eliminate v_2 in terms of ϕ and its initial value:

$$v_2(x, t) + f\phi(x, t) = v_{2_I}(x). \tag{4.71}$$

The 2-layer Lagrangian equations, thus, are

$$\ddot{X} - f\left(1 - \frac{h}{H}\right)(v_{2_I} - v_1 - f(X - x)) - \frac{1}{X'}\left[\frac{h\dot{X}^2}{H - h}\right]' + g'\left(1 - \frac{h}{H}\right)\frac{1}{X'}h' = 0, \tag{4.72}$$

$$\dot{v}_1 - \frac{\dot{X}}{1 - \frac{h}{H}}\left(\frac{v_1'}{X'} - f\frac{h}{H}\right) = 0, \tag{4.73}$$

where $h = \frac{h_I(x)}{X'}$, prime and dot denote x- and t-differentiations, respectively, and $g' = g\frac{\rho_2 - \rho_1}{\rho_2}$ – the reduced gravity in the limit $r \to 1$.

4.3.1.3 Symmetric Instability

A qualitatively new phenomena appearing in the dynamics of fronts due to the baroclinic effects is a specific symmetric instability, i.e. an instability developing without perturbations in the front-wise direction. This instability is frequently called inertial, the term "symmetric" being often reserved for its moist counterpart (e.g. Bennetts and Hoskins [2]).

For simplicity, we will consider the particular case of the initial conditions in the form of a barotropic jet with $h_{2_I} = H_2 = \text{const.}$, $v_{2_I} = v_{1_I} = v_I(x)$. By introducing the notation $\alpha_1 = \frac{H_1}{H}$, $\alpha_2 = \frac{H_2}{H}$, $\alpha_1 + \alpha_2 = 1$ we have in non-dimensional form:

$$\ddot{\phi} + \phi\frac{\alpha_1 + \phi'}{1 + \phi'} + \epsilon\frac{\alpha_1 + \phi'}{1 + \phi'}(v_1 - v_I) - \frac{\alpha_2}{1 + \phi'}\left(\frac{\dot{\phi}^2}{\alpha_1 + \phi'}\right)' - \gamma\frac{1}{(1 + \phi')^4}\phi'' = 0, \tag{4.74}$$

$$\dot{v}_1 - \frac{1}{\alpha_1 + \phi'}\dot{\phi}v_1' - \frac{\alpha_2}{\alpha_1 + \phi'}\frac{1}{\epsilon}\dot{\phi} = 0, \tag{4.75}$$

where $\epsilon = Ro = \frac{V}{fL}$ is the Rossby number based on the typical jet velocity V and typical jet width L, $\gamma = Bu = \frac{g'H\alpha_1\alpha_2}{f^2L^2}$ is the Burger number.

Equations (4.74), (4.75) are to be solved with initial conditions $\phi(x, 0) = 0$, $\dot{\phi}(x, 0) = u_{2I}$, $v_1(x, 0) = v_I$. The initial jet $v_1 = v_I$, $\phi = 0$, if non-perturbed: $u_I = 0$ is a solution.

System (4.74), (4.75) in the linear approximation gives

$$\ddot{\phi} + \alpha_1 \phi + \epsilon \alpha_1 \xi_1 - \gamma \phi'' = 0 , \tag{4.76}$$

$$\dot{\xi}_1 - \frac{v_I'}{\alpha_1} \dot{\phi} v_I' - \frac{\alpha_2}{\alpha_1} \frac{1}{\epsilon} \dot{\phi} = 0 , \tag{4.77}$$

where we introduced $v_1 - v_I = \xi$. Hence,

$$\dddot{\phi} + \dot{\phi}(1 + \epsilon v_I') - \gamma \dot{\phi}'' = 0 . \tag{4.78}$$

Using the variable $\psi = \dot{\phi}$, renormalizing x with $\sqrt{\gamma}$ and looking for the solution $\psi \propto e^{i\omega t}$, we get the quantum-mechanical Schrödinger equation:

$$\partial_{xx}^2 \psi + (E - V(x))\psi = 0 \tag{4.79}$$

for a particle having the energy $E = \omega^2$ and moving in the potential $V(x) = 1 + \epsilon v_I'$. It is worth noting that Burger number plays the role of the Planck constant squared.

It is known (e.g. Landau and Lifshits [13]) that in the case of quantum mechanical potential well there are both propagating solutions corresponding to the continuous spectrum $\omega^2 \geq 1$ and trapped in the well, localized solutions corresponding to the discrete spectrum $Min(V(x)) < \omega^2 < 1$. As is easy to see, the potential well corresponds to the region of anticyclonic shear. Hence, the trapped modes are localized there, oscillating at sub-inertial frequencies.

If the potential is deep enough (strong enough anticyclonic shear), non-oscillatory unstable modes with $\omega^2 < 0$ appear and therefore a specific instability arises. This is the symmetric instability which is thus intricately related to the presence of trapped modes inside the front. It should be noted that the known explicit solutions of the Schrödinger equations for some potentials, e.g. $\cosh^{-2}$ potential (e.g. Landau and Lifshits [13]) may be used for analytical studies of symmetric instability. The Lagrangian equations (4.74), (4.75) provide a convenient framework for studying the nonlinear stage of this instability.

4.3.1.4 Equatorial 2-Layer 1.5d RSW in Lagrangian Variables

As in the one-layer case, the Lagrangian description may be also applied to the equatorial zonal flows. The equatorial counterparts of (4.72), (4.73), with obvious interchanges between the zonal (u) and meridional (v) components of velocity and respective Lagrangian coordinates, are

$$\ddot{Y} + \beta Y u_2 + \left[-\beta Y \left((1-h)u_1 + h u_2 \right) - \left(\frac{h}{1-h} \dot{Y}^2 \right)_Y + g'(1-h) h_Y \right] = 0,$$

$$\tag{4.80}$$

$$\dot{u}_1 - \left(1 + \frac{h}{1-h} \right) \dot{Y} u_{1_Y} + \beta \frac{h}{1-h} Y \dot{Y} = 0, \tag{4.81}$$

where

$$u_2 - \beta \frac{Y^2}{2} = u_{2_I} - \beta \frac{y^2}{2}, \tag{4.82}$$

$h = \frac{h_I}{Y'}$ and $\partial_Y = \frac{1}{Y'} \partial_y$.

Introducing $\phi(y,t) = Y(y,t) - y$, linearizing around the barotropic jet $u_{2_I} = u_{1_I} = u_I$ and non-dimensionalizing as above with an obvious change for the frequency scale: $f \to \beta L$, we get the equatorial counterpart of (4.78):

$$\dddot{\phi} - \beta y (\epsilon u_I' - \alpha_2 y) \dot{\phi} - \gamma \dot{\phi}'' = 0. \tag{4.83}$$

This is an equation for linear equatorial symmetric (inertial) instability (e.g. Dunkerton [8]). Unlike the mid-latitude case, even a linear shear may lead to symmetric instability at the equator. In this case (4.83) after Fourier tranformation in t and a shift of y gives a quantum-mechanical Schrödinger equation for the harmonic oscillator with well-known solutions.

4.3.1.5 Relation to 1.5 RSW and Comments on the Pulson Solutions

A limit of strong disparity between the layers depths $\frac{h}{H} \to 0$ may be considered in (4.72), (4.73). This gives to zeroth order in $\frac{h}{H}$ a system of decoupled equations:

$$\ddot{X} - f \left(v_{2_I} - v_1 - f(X - x) \right) + g' \frac{1}{X'} \left(\frac{h_I}{X'} \right)' = 0, \tag{4.84}$$

$$\dot{v}_1 - \frac{\dot{X}}{X'} v_1' = 0. \tag{4.85}$$

We thus recover in the case of motionless upper layer, when $v_1 = 0$, the one-layer RSW equation in Lagrangian form (4.64), with the replacement $g \to g'$, which provides both a (standard) justification of the one-layer reduced-gravity model and a possibility to calculate baroclinic corrections to the one-layer RSW solutions. For example, the pulsating front solution presented in Sect. 4.2 is a zero-order in $\frac{h}{H}$ solution of (4.72), (4.73), but corrections will appear in the next orders, in particular the non-zero velocity field v_1 in the thick upper layer. They may be calculated order by order, which will be presented elsewhere. It is, however, clear that a nontrivial signature of the pulson solutions in the upper layer will appear.

4.3.2 Axisymmetric Case

As in the one-layer case, the Lagrangian approach can also be developed in the axisymmetric case. The two-layer rigid-lid RSW equations for axisymmetric configurations are described by the equations in polar coordinates r, θ:

$$(\partial_t + u_r^{(i)} \partial_r)u_r^{(i)} - u_\theta^{(i)}\left(f + \frac{u_\theta^{(i)}}{r}\right) + \partial_r \pi^{(i)} = 0, \tag{4.86a}$$

$$(\partial_t + u_r^{(i)} \partial_r)u_\theta^{(i)} + u_r^{(i)}\left(f + \frac{u_\theta^{(i)}}{r}\right) = 0, \tag{4.86b}$$

$$\partial_t h^{(i)} + \frac{1}{r}\partial_r(r\, u_r^{(i)}\, h^{(i)}) = 0, \quad i = 1, 2, \tag{4.86c}$$

$$\pi^{(1)} + g(\rho_1 h_1 + \rho_2 h_2) = \pi^{(2)}, \tag{4.86d}$$

$$h^{(1)} + h^{(2)} = 1, \tag{4.86e}$$

where $u_r^{(i)}$ and $u_\theta^{(i)}$, $i = 1, 2$ are radial and azimuthal components of the velocity, respectively, in each layer. The analog of constraint (4.58) is

$$\partial_r(rh^{(1)}u_r^{(1)} + rh^{(2)}u_r^{(2)}) = 0. \tag{4.87}$$

Hence

$$U = \frac{rh^{(1)}u_r^{(1)} + rh^{(2)}u_r^{(2)}}{H} = U(t). \tag{4.88}$$

Choosing the boundary condition of zero-radial mass flux across the vortex boundary sets $U = 0$ gives

$$u_r^{(1)} = -\frac{h^{(1)}}{H - h^{(2)}}u_r^{(2)}. \tag{4.89}$$

The pressures $\pi^{(i)}$, $i = 1, 2$ may be excluded, as in the rectilinear case, and we thus arrive at the following system of equations for four independent variables u_2, h_2, v_2 and v_1, which is the axisymmetric counterpart of (4.66), (4.67), (4.68) and (4.69):

$$\frac{\partial u_2}{\partial t} + u_2 \frac{\partial u_2}{\partial r} - \left(f + \frac{v_2}{r}\right)v_2 +$$

$$\frac{\rho_1}{\rho_2 h_1 + \rho_1 h_2}\left(\left(f + \frac{v_1}{r}\right)(h_1 v_1) + \left(f + \frac{v_2}{r}\right)(h_2 v_2) - \frac{\partial}{\partial r}\left(r(h_1 u_1^2 + h_2 u_2^2)\right) + \frac{g(\rho_2 - \rho_1)}{\rho_1}h_1 \frac{\partial h_2}{\partial r}\right) = 0, \tag{4.90}$$

$$\frac{\partial h_2}{\partial t} + \frac{\partial}{\partial r}(ru_2 h_2) = 0, \tag{4.91}$$

$$\frac{\partial v_2}{\partial t} + u_2\frac{\partial v_2}{\partial r} + u_2\left(f + \frac{v_2}{r}\right) = 0, \tag{4.92}$$

$$\frac{\partial v_1}{\partial t} + u_2\frac{\partial v_1}{\partial r} + (u_1 - u_2)\frac{\partial v_1}{\partial r} + \left(f + \frac{v_2}{r}\right)u_1 = 0, \tag{4.93}$$

where we switched back to the lower index notation for the layer number and denoted $u_r \equiv u$, $u_\theta \equiv v$.

A Lagrangian version of these equations may be easily written down along the lines of the plane-parallel case using the Lagrangian mapping $r \to R(r, t)$, the angular momentum conservation and the mass conservation $h_2 R dR = h_1 r dr$, with similar applications and conclusions, which we will not present here. It should be emphasized that centrifugal instability replaces the symmetric (inertial) instability in the axisymmetric case.

4.4 Continuously Stratified Rectilinear Fronts

4.4.1 Lagrangian Approach in the Case of Continuous Stratification

The hydrostatic primitive equations for a continuously stratified fluid with no dependence on y (the "2.5-dimensional" case) read:

$$\partial_t u + u\partial_x u + w\partial_z u - fv + g\partial_x\phi = 0, \tag{4.94a}$$

$$\partial_t v + u\partial_x v + w\partial_z v + uf = 0, \tag{4.94b}$$

$$\partial_z\phi = g\frac{\theta}{\theta_r}, \tag{4.94c}$$

$$\partial_x u + \partial_z w = 0, \tag{4.94d}$$

$$(\partial_t + u\partial_x + w\partial_z)\theta = 0. \tag{4.94e}$$

Here they are written in the atmospheric context using potential temperature θ and the so-called pseudo-height vertical coordinate (Hoskins and Bretherton [12]), θ_r is a normalization constant. For oceanic applications potential temperature should be replaced by density and the sign in the hydrostatic relation (4.94c) should be changed, z then becomes the ordinary geometric coordinate.
Potential vorticity (PV)

$$q = (f + \partial_x v)\partial_z\theta - \partial_z v\,\partial_x\theta \tag{4.95}$$

is a Lagrangian invariant $(\partial_t + u\partial_x + w\partial_z)q = 0$. As usual for straight fronts, there exist an additional Lagrangian invariant, the geostrophic momentum

$$M = v + fx,   \tag{4.96}$$

where x is understood in Lagrangian sense. The expression for the potential vorticity in terms of M is

$$q = (f + \partial_x v)\, \partial_z \theta - \partial_z v\, \partial_x \theta = \frac{\partial(M, \theta)}{\partial(x, z)}.   \tag{4.97}$$

The "slow" balanced motions are geostrophic and hydrostatic equilibria

$$u = w = 0, \quad fv = g\partial_x \phi, \quad \partial_z \phi = g\frac{\theta}{\theta_r},   \tag{4.98}$$

which are exact stationary solutions of (4.94a), (4.94b) and (4.94c) and obey the thermal wind relation

$$f\frac{\partial M}{\partial z} = \frac{g}{\theta_r}\frac{\partial \theta}{\partial x}.   \tag{4.99}$$

A potential Φ may be introduced for balanced states, such that

$$M = f^{-1}\frac{\partial \Phi}{\partial x}, \quad \theta = \frac{\theta_r}{g}\frac{\partial \Phi}{\partial z}.   \tag{4.100}$$

In fact, Φ is an "extended" geopotential given as $\Phi = \phi + f^2\frac{x^2}{2}$.
The fast motions are internal inertia gravity waves. Their dispersion relation may be easily obtained in the case of linear background stratification $\theta_0(z) = \frac{N^2}{g}\theta_r z$ by linearization about the state of rest:

$$\omega^2 = N^2\frac{k_x^2}{k_z^2} + f^2,   \tag{4.101}$$

where ω is wave frequency, $k_{x,z}$ are the wavenumber components in the horizontal and vertical directions, respectively and $N^2 = g\frac{\theta_0'(z)}{\theta_r}$.
Lagrangian variables in the vertical plane $X(x, z, t)$ and $Z(x, z, t)$ are introduced as positions at time t of the fluid particles initially found at (x, z). The incompressibility equation is written in the form of the volume conservation:

$$\frac{\partial(X, Z)}{\partial(x, z)} = 1,   \tag{4.102}$$

The primitive equations become

$$\ddot{X} + f^2 X + \frac{\partial(\phi, Z)}{\partial(x, z)} = v_I + f^2 x, \tag{4.103}$$

$$\frac{\partial(X, \phi)}{\partial(x, z)} = g \frac{\theta_I}{\theta_r}, \tag{4.104}$$

and the potential vorticity is expressed as

$$q = \frac{\partial(fx + v_I, \theta_I)}{\partial(x, z)}. \tag{4.105}$$

Elimination of ϕ by cross-differentiation gives

$$\frac{\partial(X, \ddot{X} - f v_I - f^2 x)}{\partial(x, z)} + \frac{g}{\theta_0} \frac{\partial(\theta_I, Z)}{\partial(x, z)} = 0, \tag{4.106}$$

$$\frac{\partial(X, Z)}{\partial(x, z)} = 1, \tag{4.107}$$

and for the stationary adjusted state, we get

$$\frac{\partial(X, -f v_I - f^2 x)}{\partial(x, z)} + \frac{g}{\theta_0} \frac{\partial(\theta_I, Z)}{\partial(x, z)} = 0, \tag{4.108}$$

$$\frac{\partial(X, Z)}{\partial(x, z)} = 1. \tag{4.109}$$

4.4.2 Existence and Uniqueness of the Adjusted State in the Unbounded Domain

To study the adjusted states it is convenient to use the PV equation written in terms of Φ:

$$\frac{\partial^2 \Phi}{\partial X^2} \frac{\partial^2 \Phi}{\partial Z^2} - \left(\frac{\partial^2 \Phi}{\partial X \partial Z}\right)^2 = \frac{gf}{\theta_r} q, \tag{4.110}$$

where PV in the r.h.s. is understood as a function of (X, Z). This is the Monge–Ampère equation. The boundary conditions which we will use far from the frontal zone are

$$\theta|_{z \to \pm\infty} = \theta_r \frac{N^2}{g} z, \quad N = const., \quad X|_{x \to \pm\infty} = x. \tag{4.111}$$

Although these are formally Neumann-type boundary conditions, it is easy to see that they are equivalent to the condition that far enough from the origin Φ has the form

$$\Phi|_{|X|,|Z|\to\infty} = f^2\frac{X^2}{2} + N^2\frac{Z^2}{2}. \tag{4.112}$$

This means that on some distant ellipse (which is a convex curve) $f^2\frac{X^2}{2} + N^2\frac{Z^2}{2} = $
const., the function Φ is constant, so the problem of finding the adjusted state is
reduced to the first (Dirichlet) boundary-value problem for the Monge–Ampère
equation. Existence of solution is guaranteed if the r.h.s., i.e. the PV, is continuous
and positive (Pogorelov [18]). Moreover, if the condition of convexity is added,
which is the case of (4.112), the solution is unique. Thus, for positive PV, condition
of absence of symmetric instability, the adjusted state exists and is unique in the
absence of boundaries. It is to be emphasized that the criterion is the same as for
fronts in 1- and 2-layer RSW.

An alternative Lagrangian formulation using the geostrophic and isentropic coordinates (M, θ) as independent variables in the Monge–Ampère equation was extensively used in the literature, in particular by Cullen and collaborators [5–7]. In
(M, θ) coordinates, the thermal wind relation takes the form:

$$f\frac{\partial X}{\partial \theta} = \frac{g}{\theta_r}\frac{\partial Z}{\partial M}. \tag{4.113}$$

Hence a potential Ψ for the final positions of the fluid particles may be introduced:

$$X = \frac{g}{\theta_r}\frac{\partial\Psi}{\partial M}, \quad Z = f\frac{\partial\Psi}{\partial\theta}. \tag{4.114}$$

The Jacobian of the transformation from (x, z) to (X, Z) can be rewritten as

$$\frac{\partial(X, Z)}{\partial(M, \theta)}\frac{\partial(M, \theta)}{\partial(x, z)} = 1, \tag{4.115}$$

from which we can obtain, replacing X and Z by their expressions (4.114) the following Monge–Ampère equation for Ψ with a "potential pseudo-density" which is
the inverse of the PV at the r.h.s.:

$$\frac{\partial^2\Psi}{\partial M^2}\frac{\partial^2\Psi}{\partial\theta^2} - \left(\frac{\partial^2\Psi}{\partial M\partial\theta}\right)^2 = \frac{\theta_r}{gf}\frac{1}{q}. \tag{4.116}$$

Assuming that fluid on the boundaries remains there, this equation has oblique
Neumann-type boundary conditions

$$\frac{\partial\Psi}{\partial\theta}(M_\pm(s), \theta_\pm(s)) = \frac{z_\pm}{f}, \tag{4.117a}$$

$$\left(\frac{g}{\theta_r}\frac{\partial\Psi}{\partial M}, f\frac{\partial\Psi}{\partial\theta}\right) \to (x(M, \theta), z(M, \theta)) \quad \text{as} \quad M \to \pm\infty, \tag{4.117b}$$

where $(M_\pm(s), \theta_\pm(s))$ define the upper $(z_+ = H)$ and lower boundaries $(z_- = 0)$ in (M, θ) space, s is a coordinate along those boundaries, x and z are initial positions. The domain in (M, θ) space is generally not convex which may prevent the existence of the smooth solutions of Monge–Ampère equation, although in general, this latter may be solved by methods of the optimal transportation theory (Benamou and Brenier [1]).

To illustrate the possible non-existence of the smooth solutions of the adjustment problem and the advantages of the Lagrangian approach, we give below an explicitly integrable example of zero PV in the vertically bounded domain (cf. Ou [15]). We start with a flow in a slab between $z = 0$ and $z = 1$ (in non-dimensional variables) with purely horizontal density gradients and no vertical shear in v:

$$\theta_I = \theta_I(x), \quad v_I = v_I(x), \qquad (4.118)$$

and solve (4.109). The stationary part of the horizontal momentum equation reduces to

$$\frac{\partial X}{\partial z} f(v_I' + f) + \frac{\partial Z}{\partial z} \frac{g\theta_I'}{\theta_0} = 0, \qquad (4.119)$$

where the prime denotes the x-derivative. Integration of (4.119) gives

$$X = \frac{\mathcal{F}(x)}{f v_I' + f^2} - \frac{g\theta_I'/\theta_0}{f v_I' + f^2} Z, \qquad (4.120)$$

and from the incompressibility equation it follows that

$$Z^2 \left(\frac{g\theta_I'/\theta_0}{f v_I' + f^2} \right)' - 2 \left(\frac{\mathcal{F}}{f v_I' + f^2} \right)' Z + 2(\mathcal{G}(x) + z) = 0. \qquad (4.121)$$

The functions $\mathcal{F}(x)$, $\mathcal{G}(x)$ are to be determined from the boundary conditions. For the unit strip in the x, z-plane they are

$$Z(x, 0) = 0, \, Z(x, 1) = 1. \qquad (4.122)$$

Hence

$$X = x + \mathcal{A}(x) \left(\frac{1}{2} - Z \right), \quad \mathcal{A} = \frac{g\theta_I'/\theta_0}{f v_I' + f^2}, \qquad (4.123)$$

$$Z = \frac{1}{\mathcal{A}'(x)} \left[1 + \frac{1}{2}\mathcal{A}'(x) - \sqrt{\left(1 + \frac{1}{2}\mathcal{A}'(x) \right)^2 - 2z\mathcal{A}'(x)} \right], \qquad (4.124)$$

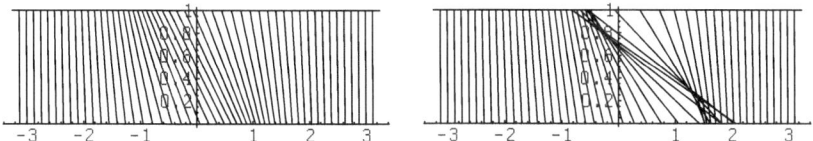

Fig. 4.5 The end-state of the evolution of the zero-PV state with initially vertical isentropic surfaces in the case of existence (*left panel*, no crossing of the isentropes) and non-existence (*right panel*, crossing of the isentropes) of the adjusted state

which is the explicit solution for the adjusted state.

If a discontinuity forms, it forms at a boundary due to the elliptic character of the problem, i.e. where $\partial_x X(x, 0) = 0$ or $\partial_x X(x, 1) = 0$. This will happen if

$$\frac{\partial X}{\partial x}(x, 0) = 1 \pm \frac{1}{2} \mathcal{A}'(x) = 0, \tag{4.125}$$

i.e. if

$$\frac{g}{f \theta_0} \left(\frac{g \theta_I' / \theta_0}{f + v_I'} \right)' = \pm 2. \tag{4.126}$$

The positions of isentropic surfaces and isotachs may be easily obtained from knowing explicit final positions of the fluid particles (4.123), (4.124) and using the Lagrangian conservation of θ and M. It may be thus shown (Plougonven and Zeitlin [17]) that singularity corresponds to intersecting isentropes, as shown in Fig. 4.5 and infinite gradients of v. We thus have a frontogenesis process, which in fact coincides with the classical scenario of Hoskins and Bretherton [12], with the only difference that in their example the parameters of the system were driven towards the singular case by an adiabatic change due to external deformation field.

4.4.3 Trapped Modes and Symmetric Instability in Continuously Stratified Case

The Lagrangian approach is also efficient for studying symmetric/inertial instability in the continuously stratified case. For this, we consider (4.94a), (4.94b), (4.94c), (4.94d) and (4.94e) taken between the flat top $z = H$ and bottom $z = 0$. We rewrite them in the form (4.107) and introduce the deviations of the particle positions from the stationary balanced state:

$$X = \bar{X} + \chi, \quad Z = \bar{Z} + \zeta, \tag{4.127}$$

so that

$$\frac{\partial(\bar{X}+\chi, \ddot{\chi})}{\partial(x, z)} - \frac{\partial(\bar{X}+\chi, f\, M_I)}{\partial(x, z)} + \frac{g}{\theta_r}\frac{\partial(\theta_I, \bar{Z}+\zeta)}{\partial(x, z)} = 0. \tag{4.128}$$

It is more convenient to use as independent variables the positions of the particles in the adjusted state $(\bar{X}, \bar{Z})$, rather than the initial positions (x, z). When this change of variables is made in (4.128), two terms which express the thermal wind relation in the adjusted state cancel out. Furthermore, it is convenient to express the gradients of $\bar{v}$ and θ in the adjusted state through the geopotential $\bar{\phi}$, making explicit use of geostrophic and hydrostatic balances. Equation (4.128) then becomes

$$\left(\frac{\partial^2}{\partial t^2} + f^2 + \frac{\partial^2\bar{\phi}}{\partial\bar{X}^2}\right)\frac{\partial\chi}{\partial\bar{Z}} + \frac{\partial^2\bar{\phi}}{\partial\bar{X}\partial\bar{Z}}\left(-\frac{\partial\chi}{\partial\bar{X}} + \frac{\partial\zeta}{\partial\bar{Z}}\right) - \frac{\partial^2\bar{\phi}}{\partial\bar{Z}^2}\frac{\partial\zeta}{\partial\bar{X}} + \frac{\partial(\chi, \ddot{\chi})}{\partial(\bar{X}, \bar{Z})} = 0. \tag{4.129}$$

The incompressibility condition gives

$$\frac{\partial\chi}{\partial\bar{X}} + \frac{\partial\zeta}{\partial\bar{Z}} + \frac{\partial(\chi, \zeta)}{\partial(\bar{X}, \bar{Z})} = 0. \tag{4.130}$$

We non-dimensionalize these equations in the context of perturbations of a balanced jet by rescaling time by f^{-1}, the horizontal displacements by U/f, where U is typical transverse velocity which is small with respect to the typical jet velocity V, so $U = \delta V$, with $\delta \ll 1$. Typical horizontal and vertical length scales are L and H, respectively. The scale of the vertical displacements is UH/fL, from the continuity equation. We thus obtain

$$\left(\frac{\partial^2}{\partial t^2} + 1 + Ro\frac{\partial^2\bar{\phi}}{\partial\bar{X}^2}\right)\frac{\partial\chi}{\partial\bar{Z}} + Ro\frac{\partial^2\bar{\phi}}{\partial\bar{X}\partial\bar{Z}}\left(-\frac{\partial\chi}{\partial\bar{X}} + \frac{\partial\zeta}{\partial\bar{Z}}\right)$$
$$- Bu\frac{\partial^2\bar{\phi}}{\partial\bar{Z}^2}\frac{\partial\zeta}{\partial\bar{X}} + +\delta\, Ro\frac{\partial(\chi, \ddot{\chi})}{\partial(\bar{X}, \bar{Z})} = 0, \tag{4.131a}$$

$$\frac{\partial\chi}{\partial\bar{X}} + \frac{\partial\zeta}{\partial\bar{Z}} + \delta\, Ro\frac{\partial(\chi, \zeta)}{\partial(\bar{X}, \bar{Z})} = 0, \tag{4.131b}$$

where $Ro = V/fL$ is the Rossby number and $Bu = N^2H^2/f^2L^2$ is the Burger number. We consider intense jets and presume $Ro \sim 1$, $Bu \sim 1$. We expand χ and ζ in small parameter δ and obtain to leading order from (4.131b):

$$\frac{\partial\chi^{(0)}}{\partial\bar{X}} + \frac{\partial\zeta^{(0)}}{\partial\bar{Z}} = 0, \tag{4.132}$$

Hence there exists a streamfunction $\psi^{(0)}$ such that

$$\chi^{(0)} = -\frac{\partial \psi^{(0)}}{\partial \bar{Z}}, \quad \zeta^{(0)} = \frac{\partial \psi^{(0)}}{\partial \bar{X}}. \qquad (4.133)$$

Equation (4.131a) becomes

$$\left(\frac{\partial^2}{\partial t^2} + 1 + \frac{\partial^2 \bar{\phi}}{\partial \bar{X}^2}\right) \frac{\partial^2 \psi^{(0)}}{\partial \bar{Z}^2} - \frac{\partial^2 \bar{\phi}}{\partial \bar{X} \partial \bar{Z}} \frac{\partial^2 \psi^{(0)}}{\partial \bar{X} \partial \bar{Z}} + \frac{\partial^2 \bar{\phi}}{\partial \bar{Z}^2} \frac{\partial^2 \psi^{(0)}}{\partial \bar{X}^2} = 0. \qquad (4.134)$$

The boundary conditions are zero vertical displacements of parcels at the top and bottom boundaries. This implies that $\psi^{(0)}$ is constant on the boundaries. As there is no overall displacement of the fluid layer in the $\bar{X}$-direction, these constants are equal; they can be both set to zero: $\psi^{(0)}(\bar{X}, 0) = \psi^{(0)}(\bar{X}, 1) = 0$. $\psi^{(0)}$ should also remain bounded as $\bar{X} \to \pm\infty$. Equation (4.134) closely resembles the homogeneous part of the Sawyer–Eliassen equation (e.g. Holton [11], p. 275) except for the term with the double-time derivative, which makes (4.134) prognostic. In the Sawyer–Eliassen equation, this term is absent because the fast time has been filtered out by balanced scaling, making the equation diagnostic.

Like in the two-layer case above, we take the simplest example of a barotropic jet. The non-dimensional geopotential describing a balanced jet is

$$\bar{\phi} = \Phi(\bar{X}) + \frac{\bar{Z}^2}{2}. \qquad (4.135)$$

The mean stratification, which we suppose constant, is given by the second derivative of the second term in this expression and the jet velocity is $\bar{v} = \Phi'$, where the prime denotes the $\bar{X}$-derivative $\Phi' = \partial \Phi / \partial \bar{X}$. If an unbalanced fast component is added to (4.135), the equation for its evolution is (cf. (4.134)):

$$(1 + \Phi'' + \partial_{tt}^2) \partial_{\bar{Z}\bar{Z}}^2 \psi^{(0)} + \partial_{\bar{X}\bar{X}}^2 \psi^{(0)} = 0. \qquad (4.136)$$

It allows a separation of variables and results in a regular Sturm–Liouville problem in the vertical results; given the boundary conditions, the vertical eigenfunctions are $\sin(n\pi \bar{Z})$. Hence the solutions are sought in the form:

$$\psi^{(0)}(\bar{X}, \bar{Z}, t) = \sum_n \sin(n\pi \bar{Z}) \, \psi_n^{(0)}(\bar{X}, t). \qquad (4.137)$$

We look for $\psi_n^{(0)}(\bar{X}, t)$ with a time dependence of the form $e^{-i\omega t}$ and get a Sturm–Liouville problem on $]-\infty, +\infty[$. We denote by $\hat{\psi}_{n\omega}(x)$ the horizontal eigenfunction with vertical wavenumber n and frequency ω. The equation for $\hat{\psi}_{n\omega}(\bar{X})$ has the form of Schrödinger equation of a particle in a well:

$$\partial_{\bar{X}\bar{X}}^2 \hat{\psi}_{n\omega} - n^2 \pi^2 (1 + \Phi'' - \omega^2) \hat{\psi}_{n\omega} = 0. \qquad (4.138)$$

The factor $n^2\pi^2$ may be removed by rescaling $\bar{X}$ as $S = n\pi\bar{X}$:

$$\partial^2_{SS}\hat{\psi}_{n\omega} - (1 + \Phi''(S/n\pi) - \omega^2)\hat{\psi}_{n\omega} = 0 , \qquad (4.139)$$

where the "potential" is $(1 + \Phi''(S/n\pi))$ and the eigenvalues are ω^2. For any given profile of Φ, the depth of the potential is always the same, but its width depends on the vertical wavenumber n: the smaller the vertical scale of the waves, the wider the potential.

The Schrödinger equation (4.139) has a continuous and a discrete spectrum of eigenvalues ω^2. The potential $(1 + \Phi'')$ tends to one as $\bar{X} \to \infty$; hence, for a given n, we have

- continuous spectrum of solutions with $\omega > 1$. This part of the spectrum is doubly degenerate (two independent solutions for each eigenvalue ω) and corresponds to leftward and rightward propagating IGW.
- discrete spectrum of solutions with subinertial frequencies:

$$\sqrt{\text{Min}(1 + \Phi'')} < \omega < 1 . \qquad (4.140)$$

This part of the spectrum is nondegenerate, and consists of solutions exponentially decaying outside the region where $(1 + \Phi'' - \omega^2) < 0$, and oscillating inside that region: they are trapped in the anticyclonic part of the jet.

- For a jet with relative vorticity lower than -1(i.e. $-f$ in dimensional variables), modes with $\omega^2 < 0$ arise in the trapped mode spectrum, yielding the symmetric instability, like in the two-layer case.

4.5 Conclusions

Thus we have shown that Lagrangian variables represent an excellent tool for handling dynamics of symmetric fronts (translational symmetry) and vortices (rotational symmetry), both qualitatively and quantitatively. Although it is known that relaxing the strict symmetry can lead to qualitatively new effects (cf., e.g. Griffiths et al. [10]), such fundamental "symmetric" phenomena as singularity formation (catastrophic adjustment) or nonlinear stage of symmetric instability are still ill-understood. We believe that the mathematical framework developed in the previous sections will allow to advance in their understanding.

References

1. Benamou, J.-D., Brenier, Y.: A computational fluid mechanics solution to the Monge–Kantorovich mass transfer problem. Numer. Math. **84**, 375–393 (2000).
2. Bennets, D.A., Hoskins, B.J.: Conditional symmetric instability – possible explanation for frontal rainbands. Q. J. R. Meteorol. Soc. **105**, 945–962 (1979).
3. Blumen, W.: Geostrophic adjustment. Rev. Geophys. Space Phys. **10**, 485–528 (1972).

4. Bouchut, F., Le Sommer, J., Zeitlin, V.: Frontal geostrophic adjustment, slow manifold and nonlinear wave phenomena in one-dimensional rotating shallow water. Part 2. Numerical simulations. J. Fluid Mech. **514**, 35–63 (2004).

5. Cullen, M.J.P., Purser, R.J.: An extended Lagrangian theory of semi-geostrophic frontogenesis. J. Atmos. Sci. **41**, 1477–1497 (1984).

6. Cullen, M.J.P., Purser, R.J.: Properties of the Lagrangian semi-geostrophic equations. J. Atmos. Sci. **46**, 2684–2697 (1989).

7. Cullen, M.J.P., Norbury, J., Purser, R.J.: Generalized Lagrangian solutions for atmospheric and oceanic flows. SIAM J. Appl. Math. **51**, 20–31 (1991).

8. Dunkerton, T.J.: On the inertial stability of the equatorial middle atmosphere. J. Atmos. Sci. **38**, 2354–2364 (1981).

9. Frei, C.: Dynamics of a two-dimensional ribbon of shallow water on an f-plane. Tellus **45A**, 44–53 (1993).

10. Griffiths, R.W., Killworth, P., Stern, M.E.: Ageostrophic instability of ocean currents. J. Fluid Mech. **117**, 343–377 (1982).

11. Holton, J.R.: An introduction to dynamic meteorology. Academic, San Diego (1992).

12. Hoskins, B.J., Bretherton, F.P.: Atmospheric frontogenesis models: mathematical formulation and solution. J. Atmos. Sci. **29**, 11–37 (1972).

13. Landau, L.D., Lifshits, E.M.: Quantum Mechanics, Academic, New York (1975).

14. LeSommer, J., Medvedev, S.B., Plougonven, R., Zeitlin, V.: Singularity formation during the relaxation of jets and fronts towards the state of geostrophic equilibrium. Commun. Nonlinear Sci. Numer. Simul. **8**, 415–442 (2003).

15. Ou, H.W.: Geostrophic adjustment: a mechanism for frontogenesis. J. Phys. Oceanogr. **14**, 994–1000 (1984).

16. Pedlosky, J.: Geophysical Fluid Dynamics, Springer, New York (1982).

17. Plougonven, R., Zeitlin, V.: Lagrangian approach to geostrophic adjustment of frontal anomalies in stratified fluid. Geophys. Astrophys. Fluid Dyn. **99**, 101–135 (2005).

18. Pogorelov, A.V.: Extrinsic Geometry of Convex Surfaces. AMS, Providence (1973).

19. Reznik, G.M., Zeitlin, V.: Asymptotic methods with applications to the fast–slow splitting and the geostrophic adjustment. In: Zeitlin, V. (ed.) Nonlinear Dynamics of Rotating Shallow Water. Methods and Advances, pp. 47–120. Elsevier, Amsterdam (2007).

20. Reznik, G.M., Zeitlin, V., Ben Jelloul, M.: Nonlinear theory of geostrophic adjustment. Part I. Rotating shallow water. J. Fluid Mech. **445**, 93–120 (2001).

21. Rubino, A., Brandt, P., Hessner, K.: Analytic solutions for circular eddies of the reduced-gravity shallow-water equations. J. Phys. Oceanogr. **28**, 999–1002 (1998).

22. Rubino, A., Dotsenko, S., Brandt, P.: Near-inertial oscillations of geophysical surface frontal currents. J. Phys. Oceanogr. **33**, 1990–1999 (2003).

23. Sutyrin, G., Zeitlin, V.: Generation of inertia-gravity waves by pulsating lens-like axisymmetric vortices. In: Proceedings of the 18th French Mechanical Congress, Grenoble, CFM2007-0806. http://hdl.handle.net/2042/15595 (2007).

24. Zeitlin, V.: Nonlinear wave phenomena in rotating shallow water with applications to geostrophic adjustment. In: Zeitlin, V. (ed.) Nonlinear Dynamics of Rotating Shallow Water. Methods and Advances, pp. 257–322. Elsevier, Amsterdam (2007).

25. Zeitlin, V., Medvedev, S.B., Plougonven, R.: Frontal geostrophic adjustment, slow manifold and nonlinear wave phenomena in one-dimensional rotating shallow water. Part 1. Theory. J. Fluid Mech. **481**, 269–290 (2003).

Chapter 5
Wave–Vortex Interactions

O. Bühler

This chapter presents a theoretical investigation of wave–vortex interactions in fluid systems of interest to atmosphere and ocean dynamics. The focus is on *strong* interactions in the sense that the induced changes in the vortical flow should be significant. In essence, such strong wave–vortex interactions require significant changes in the potential vorticity (PV) of the flow either by advection of pre-existing PV contours or by creating new PV structures via wave dissipation and breaking. This chapter explores the interplay between wave and PV dynamics from a theoretical point of view based on a recently formulated conservation law for the sum of mean-flow impulse and wave pseudomomentum.

First, the conservation law is derived using elements of generalized Lagrangian mean theory such as the Lagrangian definition of pseudomomentum. Then the creation of vorticity due to breaking and dissipating waves is explored using the shallow water system and the example of wave-driven longshore currents and vortices on beaches, especially beaches with non-trivial topography. This is followed by an investigation of wave refraction by vortices and the concomitant back reaction on the vortices both in shallow water and in three-dimensional stratified flow.

Particular attention is paid to the phenomenon of wave capture in three dimensions and to the peculiar duality between wavepackets and vortex couples that it entails.

5.1 Introduction

We are interested in the nonlinear interactions between waves and vortices in fluid systems such as the two-dimensional shallow water system or the three-dimensional Boussinesq system. In particular, we concentrate on waves whose dynamics has no essential dependence on potential vorticity (PV), so a typical example would be surface gravity waves in shallow water (or internal gravity waves

O. Bühler (✉)
Courant Institute of Mathematical Sciences, New York University,
251 Mercer Street, New York, NY 10012, USA, obuhler@cims.nyu.edu

Bühler, O.: *Wave–Vortex Interactions*. Lect. Notes Phys. **805**, 139–187 (2010)
DOI 10.1007/978-3-642-11587-5_5

in three-dimensional stratified flow) and their interactions with the layerwise two-dimensional vortices familiar from quasi-geostrophic dynamics.

Many such interactions are possible, but we focus on *strong* interactions, which are defined by their capacity to lead to significant $O(1)$ changes of the PV field even for small-amplitude waves. More specifically, if the wave amplitude is given by a non-dimensional parameter $a \ll 1$ and if the governing equations are expanded in powers of a, then the linear wave dynamics occurs at $O(a)$ and the leading-order nonlinear interactions occur at $O(a^2)$. A strong interaction occurs if the wave-induced $O(a^2)$ changes in the PV can grow secularly in time such that over long times $t = O(a^{-2})$ these PV changes may accrue to be $O(1)$. Naturally, this involves some kind of resonance of the wave-induced forcing terms with the PV-controlled linear mode in order to achieve the secular growth $O(a^2 t)$ in the PV changes.

This straightforward perturbation expansion in small wave amplitude easily obscures an all-important physical fact that is not restricted to small wave amplitudes. It is clear from fundamental fluid dynamics that strong interactions between waves and vortices require the achievement of significant wave-induced changes in the potential vorticity (PV) distribution of the flow. However, such changes are tightly constrained by the material invariance of the potential vorticity in perfect fluid flow, which is a consequence of Kelvin's circulation theorem. As an example, consider the standard one-layer shallow-water equations with Cartesian coordinates $x = (x, y)$, velocity components $u = (u, v)$, and layer depth h such that

$$\frac{Du}{Dt} + g\nabla(h - H) = F \quad \text{and} \quad \frac{Dh}{Dt} + h\nabla \cdot u = 0. \tag{5.1}$$

Here $D/Dt = \partial_t + u \cdot \nabla$ is the material derivative, g is gravity, F is some body force, and $H(x)$ is the possibly non-uniform still water depth such that $h - H$ is the surface elevation (see Fig. 5.1). The potential vorticity is given by

$$q = \frac{\nabla \times u}{h} \quad \text{such that} \quad \frac{Dq}{Dt} = \frac{\nabla \times F}{h}, \tag{5.2}$$

where $\nabla \times u = v_x - u_y$.

Now, the point is that for perfect fluid flow $F = 0$ and therefore q is a material invariant. This makes obvious that for perfect fluid flow any changes in the spatial

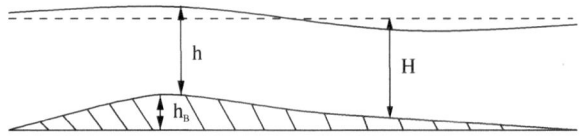

Fig. 5.1 Shallow-water layer with still water depth H and topography h_B. For non-uniform bottom topography $h - H$ is the surface elevation. In the case of uniform bottom topography the still water depth is constant and can be ignored

distribution of q must be due to advection of fluid particles across a pre-existing PV gradient. Strong interactions between gravity waves and vortices are possible only if the gravity waves can lead to large $O(1)$ displacements of fluid particles in the direction of the PV gradient. Examples of this kind of non-dissipative scenario do exist [e.g. 15, 17], but more commonly observed is the lack of strong interactions between waves and vortices in perfect fluid flow. This is essentially due to the resilience of circular vortices to large irreversible deformations.[1]

This indicates the importance of non-perfect flow effects for strong wave–vortex interactions. Perhaps the most important such effect is wave dissipation, which leads to $F \neq 0$ and therefore to material changes in the PV. Wave dissipation can be due either to laminar viscous effects or due to nonlinear wave breaking and the concomitant breakdown of the organized wave motion into three-dimensional turbulence, as exemplified by the breaking of surface waves on a beach. We will take the view that both forms of wave dissipation can be treated on the same footing as far as the wave–vortex interactions are concerned. Consideration of the wave-induced changes in PV due to dissipating waves leads to the well-known phenomenon of *wave drag* which is the standard term for the effective mean force exerted on the mean flow due to steady but dissipative waves.[2]

For instance, wave drag is central for the generation of longshore currents by obliquely incident surface waves on a beach, for the reduced speed of the high-altitude mesospheric jet in the atmosphere due to dissipating topographic waves, and for the maintenance and shape of the global circulation of the middle atmosphere [e.g. 35]. The situation is less clear in the deep ocean, where wave drag seems to be less important than the small-scale mixing induced by the breaking waves [43].

We will look at both dissipative and non-dissipative wave–vortex interactions in this chapter. A useful theoretical tool is the definition of the Lagrangian mean velocity and of the *pseudomomentum vector* as they were introduced in the generalized Lagrangian mean GLM theory of Andrews and McIntyre [2, 3]. These Lagrangian (i.e. particle-following) definitions allow writing down a circulation theorem and corresponding PV evolution for the *Lagrangian mean flow* as defined by a suitable averaging procedure. In contrast, this does not work for the Eulerian mean flow. In this chapter we consider small-scale waves and large-scale vortices, so there is a natural scale separation that can be used for averaging. This is the standard averaging over the rapidly varying phase of a wavetrain whose amplitude and central wavenumber vary slowly in space and time. Another advantage is that in this regime

---

[1] A special case is one-dimensional shallow-water flow, in which significant and irreversible material deformations are ruled out a priori. In this case there are no strong wave–vortex interactions in perfect fluid flow [29].

[2] The connection between wave drag and PV dynamics is somewhat obscured in the standard treatments of this phenomenon, which are based on zonally symmetric mean flows [e.g. 1].

there are simple relations between Lagrangian and the more familiar Eulerian mean quantities. For instance, we shall see that in shallow water the pseudomomentum, Stokes drift, and bolus velocity (i.e. the eddy-induced transport velocity) are all approximately equal in this regime.

Now, the main theoretical result is a conservation law for the sum of the total pseudomomentum and the impulse of the mean PV field, with impulse to be defined below. This conservation law expresses a certain wave–vortex duality, which allows understanding the essence of various interactions even without detailed computations, which is a distinct practical advantage. Examples are given for the dissipative generation of PV by breaking shallow-water waves and for the non-dissipative refraction of waves by vortical mean flows, which can lead to irreversible scattering of the waves. The latter leads to a peculiar irreversible feedback on the PV structure termed *remote recoil* in [16], which is very well explained by the aforementioned conservation law. The same effect is even stronger for internal gravity waves in the three-dimensional Boussinesq system, where refraction can lead to a peculiar form of non-dissipative wave destruction termed *wave glueing* or *wave capture*, which is due to the advection and straining of wave phase by the vortical mean flow [4, 17].

All these examples serve to illustrate the interplay between PV evolution and the dynamics of the waves and how strong interactions are compatible with constraints on PV dynamics that follow from the exact PV evolution law (5.2). The plan of this chapter is as follows. In Sect. 5.2 the Lagrangian mean flow and pseudomomentum are introduced, the mean circulation theorem is written down, and the simple relations between various Lagrangian and Eulerian quantities in the regime of a slowly varying wavetrain are noted. This leads to the conservation law for pseudomomentum and impulse. In Sect. 5.3 the PV generation by breaking waves in shallow water is discussed and its application to vortex dynamics on beaches is described in Sect. 5.4. Refraction of waves by the vortical mean flow and the attendant wave–vortex interactions are discussed in Sect. 5.5 both in shallow water and in the three-dimensional Boussinesq system. Finally, concluding comments are offered in Sect. 5.6.

5.2 Lagrangian Mean Flow and Pseudomomentum

Here we introduce the elements of GLM theory that are most useful for studying wave–vortex interactions. GLM theory is described in full in [2, 3] and more detailed introductions to some of the elements used here can be found in [15, 11] and in the forthcoming book [13]. The effort to understand these elements of GLM theory is not very great and they provide very useful reference points for the interaction dynamics. Overall, the aim is not to present a full set of GLM equations, but rather to extract a minimal set of equations that captures most of the constraints that Kelvin's circulation theorem puts on wave–vortex interactions. We focus on the two-dimensional shallow-water system, but this material readily generalizes to three-dimensional flow (e.g. [17]).

5.2.1 Lagrangian Averaging

GLM theory is based on two elements: an Eulerian averaging operator $\overline{(\ldots)}$ and a disturbance-associated particle displacement field $\boldsymbol{\xi}(\boldsymbol{x}, t)$. Averaging allows writing any flow field ϕ as the sum of a mean and a disturbance part $\phi = \overline{\phi} + \phi'$, say. The choice of the averaging operator is quite arbitrary provided it has the projection property $\overline{\phi'} = 0$, which makes the flow decomposition unique. For instance, zonal averaging for periodic flows is a common averaging operator in atmospheric fluid dynamics.

In our case averaging means phase averaging over the rapidly varying phase of the wavetrain, which can also be thought of as time averaging over the high-frequency oscillation of the waves. More specifically, if the oscillations are rapid enough, then one can distinguish between the evolution on the "fast" timescale of the oscillations and the evolution on the "slow" timescale of the remaining fields such as the wavetrain amplitude. This could be made explicit by introducing multiple timescales such that t/ϵ is the fast time for $\epsilon \ll 1$, for instance. We will suppress this extra notation and leave it understood that $\boldsymbol{\xi}$ and the other disturbance fields are evolving on fast and slow timescales whereas $\overline{\boldsymbol{u}}^L$ evolves on the slow timescale only.

The new field $\boldsymbol{\xi}$ is easily visualized in the case of a timescale separation (see Fig. 5.2): the location $\boldsymbol{x} + \boldsymbol{\xi}(\boldsymbol{x}, t)$ is the *actual* position of the fluid particle whose *mean* (i.e. time-averaged) position is $\boldsymbol{x}$ at (slow) time t. This goes together with $\overline{\boldsymbol{\xi}} = 0$, i.e. $\boldsymbol{\xi}$ has no mean part by definition. This definition of $\boldsymbol{\xi}$ is a natural extension of the usual small-amplitude particle displacements often used in linear wave theory. With $\boldsymbol{\xi}$ in hand we can define the Lagrangian mean of any flow field as

$$\overline{\phi}^{L} = \overline{\phi(\boldsymbol{x} + \boldsymbol{\xi}(\boldsymbol{x}, t), t)}, \tag{5.3}$$

where the opulent notation makes explicit where $\boldsymbol{\xi}$ is evaluated. From now we resolve that we will never evaluate $\boldsymbol{\xi}$ anywhere else but at $\boldsymbol{x}$ and t, so we can omit its arguments henceforth.

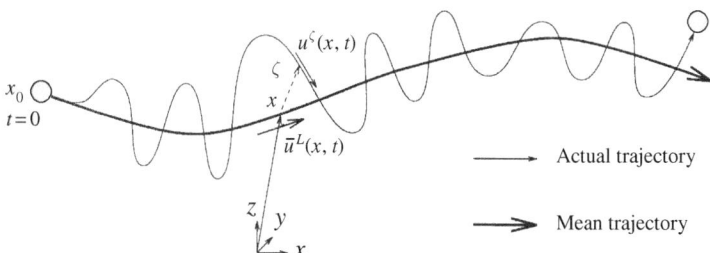

Fig. 5.2 Mean and actual trajectories of a particle in problem with multiple timescales: $\boldsymbol{x} + \boldsymbol{\xi}(\boldsymbol{x}, t)$ is the *actual* position of the fluid particle whose *mean* position is $\boldsymbol{x}$ at (slow) time t. The notation $\boldsymbol{u}^{\xi}(\boldsymbol{x}, t)$ is shorthand for $\boldsymbol{u}(\boldsymbol{x} + \boldsymbol{\xi}(\boldsymbol{x}, t), t)$

Now, by construction (5.3) constitutes a Lagrangian average over fixed particles rather than a Eulerian average over a fixed set of positions. To round off the kinematics of GLM theory we note that it can be shown that

$$\overline{D}^L(x + \xi) = u(x + \xi, t) \quad \Rightarrow \quad \overline{D}^L \xi = u(x + \xi, t) - \overline{u}^L(x, t) \qquad (5.4)$$

where $\overline{D}^L = \partial_t + \overline{u}^L \cdot \nabla$ is the Lagrangian mean material derivative. This ensures that $x + \xi$ moves with the actual velocity if x moves with the mean velocity $\overline{u}^L$.

The main motivation to work with Lagrangian mean quantities lies in the following formula:

$$\frac{D\phi}{Dt} = S \quad \Rightarrow \quad \overline{\frac{D\phi}{Dt}}^L = \overline{D}^L \overline{\phi}^L = \overline{S}^L. \qquad (5.5)$$

In particular, if the source term $S = 0$, then ϕ is a material invariant and $\overline{\phi}^L$ is a Lagrangian mean material invariant, i.e. $\overline{\phi}^L$ is constant along trajectories of the Lagrangian mean velocity $\overline{u}^L$. Again, such simple kinematic results are not available for the Eulerian mean $\overline{\phi}$, which evolves according to

$$(\partial_t + \overline{u} \cdot \nabla)\overline{\phi} = \overline{S} - \overline{u' \cdot \nabla \phi'}. \qquad (5.6)$$

This illustrates the loss of Lagrangian conservation laws that is typical for Eulerian mean flow theories.

In general, $\overline{\phi}^L \neq \overline{\phi}$ and the difference is referred to as the Stokes correction or Stokes drift in the case of velocity, i.e.

$$\overline{\phi}^L = \overline{\phi} + \overline{\phi}^S. \qquad (5.7)$$

For small-amplitude waves $\xi = O(a)$ and then the leading-order Stokes correction can be found from Taylor expansion as

$$\overline{\phi}^S = \overline{\xi_j \phi'_{,j}} + \frac{1}{2}\overline{\xi_i \xi_j} \overline{\phi}_{,ij} + O(a^3), \qquad (5.8)$$

where index notation is with summation over repeated indices understood. The first term dominates if mean flow gradients are weak.

5.2.2 Pseudomomentum and the Circulation Theorem

The circulation Γ around a closed material loop $\mathcal{C}^\xi$, say, is defined in a two-dimensional domain by

$$\Gamma = \oint_{\mathcal{C}^\xi} \boldsymbol{u}(\boldsymbol{x}, t) \cdot d\boldsymbol{x} = \int_{\mathcal{A}^\xi} \nabla \times \boldsymbol{u} \, dx dy. \tag{5.9}$$

The second form uses Stokes's theorem and $\mathcal{A}^\xi$ is the area enclosed by $\mathcal{C}^\xi$, i.e. $\mathcal{C}^\xi = \partial \mathcal{A}^\xi$. As written, the material loop $\mathcal{C}^\xi$ is formed by the *actual* positions of a certain set of fluid particles. Under the assumption[3] that the map

$$\boldsymbol{x} \to \boldsymbol{x} + \boldsymbol{\xi} \tag{5.10}$$

is smooth and invertible, we can associate with each such actual position also a *mean* position of the respective particle, and the set of all mean positions then forms another closed loop $\mathcal{C}$, say. In other words, we define the mean loop $\mathcal{C}$ via

$$\boldsymbol{x} \in \mathcal{C} \quad \Leftrightarrow \quad \boldsymbol{x} + \boldsymbol{\xi}(\boldsymbol{x}, t) \in \mathcal{C}^\xi. \tag{5.11}$$

This allows rewriting the contour integral in (5.9) in terms of $\mathcal{C}$, which mathematically amounts to a variable substitution in the integrand. The only non-trivial step is the transformation of the line element $d\boldsymbol{x}$, which is

$$d\boldsymbol{x} \to d(\boldsymbol{x} + \boldsymbol{\xi}) = d\boldsymbol{x} + (d\boldsymbol{x} \cdot \nabla)\boldsymbol{\xi}. \tag{5.12}$$

In index notation this corresponds to

$$dx_i \to dx_i + \xi_{i,j} \, dx_j. \tag{5.13}$$

This leads to

$$\Gamma = \oint_{\mathcal{C}} (u_i(\boldsymbol{x} + \boldsymbol{\xi}, t) + \xi_{j,i} u_j(\boldsymbol{x} + \boldsymbol{\xi}, t)) \, dx_i \tag{5.14}$$

after renaming the dummy indices. The integration domain is now a mean material loop and therefore we can average (5.14) by simply averaging the factors multiplying the mean line element $d\boldsymbol{x}$. The first term brings in the Lagrangian mean velocity and the second term serves as the definition of the pseudomomentum, i.e.

$$\overline{\Gamma} = \oint_{\mathcal{C}} (\overline{\boldsymbol{u}}^L - \mathbf{p}) \cdot d\boldsymbol{x} \quad \text{where} \quad \mathsf{p}_i = -\overline{\xi_{j,i} u_j(\boldsymbol{x} + \boldsymbol{\xi}, t)} \tag{5.15}$$

is the GLM definition of the pseudomomentum vector; the minus sign is conventional and turns out to be convenient in wave applications. This exact kinematic relation shows that the mean circulation is due to a cooperation of $\overline{\boldsymbol{u}}^L$ and $\mathbf{p}$, i.e. both the mean flow and the wave-related pseudomomentum contribute to the circulation.

---

[3] This can fail for large waves.

In perfect fluid flow the circulation is conserved by Kelvin's theorem and hence $\overline{\Gamma} = \Gamma$. Just as Γ is constant because C^ξ follows the actual fluid flow we now also have that $\overline{\Gamma}$ is constant because C follows the Lagrangian mean flow. This mean circulation conservation statement alone has powerful consequences if the flow is zonally periodic and the Eulerian-averaging operation consists of zonal averaging, which is the typical setup in atmospheric wave–mean interaction theory. In this periodic case a material line traversing the domain in the zonal x-direction qualifies as a closed loop for Kelvin's circulation theorem. By construction, $\partial_x \overline{(\ldots)} = 0$ for any mean field, and therefore a straight line in the zonal direction qualifies as a mean closed loop. The mean conservation theorem then implies theorem I of [2], i.e.

$$\overline{D}^L \overline{u}^L = \overline{D}^L \mathsf{p}_1, \tag{5.16}$$

where p_1 is the zonal component of $\mathbf{p}$. This is an exact statement and its straightforward extension to forced–dissipative flows constitutes the most general statement about so-called non-acceleration conditions, i.e. wave conditions under which the zonal mean flow is not accelerated. These are powerful statements, but their validity is restricted to the simple geometry of periodic flows combined with zonal averaging.

In order to exploit the mean form of Kelvin's circulation theorem for more general flows, we need to derive its local counterpart in terms of vorticity or potential vorticity. Indeed, the mean circulation theorem implies a mean material conservation law for a mean PV by the same standard construction that yields (5.2) from Kelvin's circulation theorem. Specifically, the invariance of Γ in the second form in (5.9) for arbitrary infinitesimally small material areas $\mathcal{A}^\xi$ implies the material invariance of $\nabla \times \boldsymbol{u}\, dxdy$. The area element $dxdy$ is not a material invariant in compressible shallow-water flow, but the mass element $h\, dxdy$ is. Factorizing with h leads to

$$\frac{D}{Dt}\left(\frac{\nabla \times \boldsymbol{u}}{h} h\, dxdy\right) = 0 \quad \Rightarrow \quad \frac{D}{Dt}\left(\frac{\nabla \times \boldsymbol{u}}{h}\right) = 0, \tag{5.17}$$

which is (5.2) for perfect flow. Mutatis mutandis, the same argument applied to (5.15) yields

$$\overline{q}^L = \frac{\nabla \times (\overline{u}^L - \mathbf{p})}{\tilde{h}} \quad \text{and} \quad \overline{D}^L \overline{q}^L = 0, \tag{5.18}$$

provided the mean layer depth $\tilde{h}$ is defined such that $\tilde{h}\, dxdy$ is the mean mass element, which is invariant following $\overline{u}^L$. This is true if $\tilde{h}$ satisfies the mean continuity equation

$$\overline{D}^L \tilde{h} + \tilde{h}\nabla \cdot \overline{u}^L = 0. \tag{5.19}$$

Unfortunately, $\tilde{h} \neq \overline{h}^L$ in general, which is a disadvantage of GLM theory. It can be shown that $\tilde{h}(\boldsymbol{x}, t) = h(\boldsymbol{x} + \boldsymbol{\xi}, t) J(\boldsymbol{x}, t)$, where $J = \det(\delta_{ij} + \xi_{i,j})$ is the Jacobian of the map (5.10).

The mean circulation theorem is an exact statement, so in particular it is not limited to small wave amplitudes. It shows that the Lagrangian mean flow inherits a version of the constraints that Kelvin's circulation theorem puts on strong wave–vortex interactions. For example, in irrotational flows we have $q = 0$ and therefore

$$\overline{q}^L = 0 \quad \Rightarrow \quad \nabla \times \overline{\boldsymbol{u}}^L = \nabla \times \mathbf{p}. \tag{5.20}$$

This shows that if $\nabla \times \mathbf{p}$ is uniformly bounded at $O(a^2)$ in time then so is $\nabla \times \overline{\boldsymbol{u}}^L$, which rules out strong interactions based on mean flow vorticity. Of course, even though $\overline{\boldsymbol{u}}^L$ and $\mathbf{p}$ have the same curl they can still be different vector fields. This can be either because $\nabla \cdot \overline{\boldsymbol{u}}^L$ is markedly different from $\nabla \cdot \mathbf{p}$ or because $\overline{\boldsymbol{u}}^L$ and $\mathbf{p}$ satisfy different boundary conditions at impermeable walls (see [2] for an example involving sound waves). Any strong wave–vortex interaction in the present case of irrotational flow must therefore involve wavelike behaviour of the mean flow itself, with significant values of $\nabla \cdot \overline{\boldsymbol{u}}^L$ for instance.

If $q \neq 0$, then (5.20) is replaced by

$$\nabla \times \overline{\boldsymbol{u}}^L = \nabla \times \mathbf{p} + \tilde{h}\overline{q}^L. \tag{5.21}$$

This illustrates the scope for further changes in $\nabla \times \overline{\boldsymbol{u}}^L$ due to dilation effects mediated by variable $\tilde{h}$ (i.e. vortex stretching) or due to material advection of different values of $\overline{q}^L$ into the region of interest. The latter process requires the existence of a PV gradient, as discussed earlier. Obviously, any knowledge of bounds on changes in $\tilde{h}$ and $\overline{q}^L$ can be converted into bounds on changes in $\nabla \times \overline{\boldsymbol{u}}^L$ by using the exact (5.21) as a constraint.

5.2.3 Impulse Budget of the GLM Equations

The impulse (also called Kelvin's impulse or hydrodynamical impulse) is a classical concept in incompressible constant-density fluid dynamics going back to Kelvin [e.g. 30, 6]. In essence, the impulse complements the standard momentum budget whilst being based strictly on the vorticity of the flow. This can be a very powerful tool. We start by describing the classical impulse concept and then we go on to define a useful impulse for the GLM equations.

The classical impulse is a vector-valued linear functional of the vorticity defined by

$$\text{impulse} \quad = \quad \frac{1}{n-1} \int \boldsymbol{x} \times (\nabla \times \boldsymbol{u}) \, dV, \tag{5.22}$$

where $n > 1$ is the number of spatial dimensions, dV is the area or volume element, and the integral is extended over the flow domain. We are most interested in the two-dimensional case, in which

$$\text{two-dimensional impulse} \quad = \int (y, -x) \, \nabla \times \boldsymbol{u} \, dx dy. \qquad (5.23)$$

The impulse has a number of remarkable properties for incompressible perfect fluid flow. To begin with, the impulse is clearly well defined whenever the vorticity is compact, i.e. whenever the vorticity has compact support such that $\nabla \times \boldsymbol{u} = 0$ outside some finite region. If $n = 3$ then this fixes the impulse uniquely, but if $n = 2$ then the value of the impulse depends on the location of the coordinate origin unless the net integral of $\nabla \times \boldsymbol{u}$, which is the total circulation around the fluid domain, is zero. For example, in two dimensions the impulse of a single point vortex with circulation Γ is equal to $\Gamma(Y, -X)$ where (X, Y) is the position of the vortex. This illustrates the dependence on the coordinate origin. On the other hand, two point vortices with equal and opposite circulations $\pm\Gamma$ separated by a distance d yield a coordinate-independent impulse vector with magnitude Γd and direction parallel to the propagation direction of the vortex couple. To fix this image in your mind you can consider the impulse of the trailing vortices behind a tea (or coffee) spoon: the impulse is always parallel to the direction of the spoon motion.

The easily evaluated impulse integral in an unbounded domain contrasts with the momentum integral, which in the same situation is not absolutely convergent and therefore is not well defined [30, 40, 12]. For instance, in the case of the two-dimensional vortex couple the velocity field decays as $1/r^2$ with distance r from the couple, which is not fast enough to make the momentum integral absolutely convergent. Thus a vortex couple in an unbounded domain has a unique impulse, but no unique momentum.

As far as dynamics is concerned, it can be shown that the unforced incompressible Euler equations in an unbounded domain conserve the impulse. The proof involves time-differentiating (5.22) and using integration by parts together with an estimate of the decay rate of $\boldsymbol{u}$ in the case of a compact vorticity field. Moreover, if the flow is forced by a body force $\boldsymbol{F}$ with compact support, then the time rate of change of the impulse is equal to the net integral of $\boldsymbol{F}$. This follows from the vorticity equation in conjunction with a useful integration-by-parts identity for arbitrary vector fields with compact support:

$$\int \boldsymbol{F} \, dV = -\int \boldsymbol{x} \, \nabla \cdot \boldsymbol{F} \, dV = \frac{1}{n-1} \int \boldsymbol{x} \times (\nabla \times \boldsymbol{F}) \, dV. \qquad (5.24)$$

The integrals are extended over the support of $\boldsymbol{F}$ and the second term is included for completeness; it illustrates that $\nabla \cdot \boldsymbol{F}$ and $\nabla \times \boldsymbol{F}$ are not independent for compact vector fields. Note that (5.24) does not apply to the velocity $\boldsymbol{u}$ because $\boldsymbol{u}$ does not have compact support. Now, in the tea spoon example the impulse of the trailing

vortex couple can be equated to the net force exerted by the spoon.[4] This illustrates
how impulse concepts are useful for fluid–body interaction problems. For example,
similar impulse concepts have been used to study the bio-locomotion of fish [19]
and of water-walking insects [12].

In a bounded domain the situation is somewhat different. Now the momentum
integral for incompressible flow is convergent and in fact the net momentum is
exactly zero because the centre of mass of an enclosed body of homogeneous fluid
cannot move. The impulse, on the other hand, is nonzero and usually not constant in
time anymore. This is obvious by considering the example of a vortex couple propagating towards a wall, which increases the separation d of the vortices and thereby
increases[5] impulse. However, the instantaneous rate of change of the impulse due
to a compact body force F is still given by the net integral of F. This works best
if F is large but applies only for a short time interval, because then the boundary-
related impulse changes are negligible during this short interval. Indeed, this kind
of "impulsively forced" scenario gave the impulse its name. Finally, intermediate
cases such as a zonal channel geometry are also possible, in which the flow domain
is periodic or unbounded in x, but is bounded by two parallel straight walls in y. In
this case the x-component of impulse is still exactly conserved under unforced flow,
but not the y-component.

So now the question is whether the impulse concept can be applied to wave–
vortex interactions. The idea is to define a suitable mean flow impulse that evolves
in a useful way under such interactions. This raises two issues. First, the classical impulse concept is restricted to incompressible flow, i.e., if compressible flow
effects are allowed, then most of the useful conservation properties of the impulse
are lost. Still, the vortical mean flow dynamics, especially in the geophysically relevant regime of slow layer-wise two-dimensional flow, is often characterized by
weak two-dimensional compressibility; a case in point is standard quasi-geostrophic
dynamics in which the horizontal divergence is negligible at leading order. This
suggests that two-dimensional impulse may still be useful. The second issue is the
question as to which velocity field to use to form the impulse as in (5.23). For
instance, one could base the GLM impulse on $\overline{u}^L$, but it turns out to be much more
convenient to base the GLM impulse on $\overline{u}^L - \mathbf{p}$ instead [17]. We therefore define
the GLM impulse in the shallow water system as

$$\mathcal{I} = \int (y, -x)\, \nabla \times (\overline{u}^L - \mathbf{p})\, dx\, dy = \int (y, -x)\, \overline{q}^L \tilde{h}\, dx\, dy, \qquad (5.25)$$

where the integral extends over the flow domain, as before. Clearly, $\mathcal{I}$ is well defined
if $\overline{q}^L$ has compact support, which is a property that can be controlled from the initial

---

[4] More precisely, the time rate of change of the impulse equals the instantaneous force exerted by
the spoon; time-integration then yields the final answer.

[5] It is a counter-intuitive fact that as d increases the impulse of the vortex couple increases even
though its propagation velocity decreases! Indeed, the impulse is proportional to d and the velocity
to $1/d$.

conditions of the flow together with the mean material invariance of $\overline{q}^L$. Also, $\mathcal{I}$ is obviously zero in the case of irrotational flow. This suggest that $\mathcal{I}$ is targeted on the vortical part of the flow, which is what we want, but the important question is how $\mathcal{I}$ evolves in time. The easiest way to find the time derivative of $\mathcal{I}$ in the case of compact $\overline{q}^L$ is by interpreting the integral in (5.25) as an integral over a material area that is strictly larger than the support of $\overline{q}^L$. The time derivative of such a material integral can then be evaluated by applying $\overline{D}^L$ to the entire integrand, including $dxdy$. However, as both $\overline{q}^L$ and $\tilde{h}dxdy$ are mean material invariants the only nonzero term comes from $\overline{D}^L(y, -x) = (\overline{v}^L, -\overline{u}^L)$. After some integration by parts this yields

$$\frac{d\mathcal{I}}{dt} = \int (\overline{u}^L - \mathbf{p})\,\nabla\cdot\overline{u}^L\,dxdy + \int (\nabla\overline{u}^L)\cdot\mathbf{p}\,dxdy + \text{remainder}. \qquad (5.26)$$

Here the $\mathbf{p}$ contracts with $\overline{u}^L$ and not with ∇, i.e. in index notation the second integrand is $\overline{u}^L_{j,i}\mathsf{p}_j$ with free index i. Explicitly,

$$(\nabla\overline{u}^L)\cdot\mathbf{p} = (\overline{u}^L_x\mathsf{p}_1 + \overline{v}^L_x\mathsf{p}_2, \overline{u}^L_y\mathsf{p}_1 + \overline{v}^L_y\mathsf{p}_2) \qquad (5.27)$$

in terms of the pseudomomentum components $\mathbf{p} = (\mathsf{p}_1, \mathsf{p}_2)$.

The remainder in (5.26) consists of integrals over derivatives such as $\overline{v}^L_x\overline{v}^L = 0.5\partial_x(\overline{v}^L)^2$ or $(\overline{v}^L\mathsf{p}_2)_x$, which yield vanishing contributions in an unbounded domain if $\overline{u}^L$ and $\mathbf{p}$ decay fast enough with distance r. For example, a decay $\overline{u}^L = O(1/r)$ or $\overline{u}^L = O(1/r^2)$ is sufficient, respectively, depending on whether $\mathbf{p}$ is compact or not. We will assume that $\mathbf{p}$ is compact in our examples (unless an explicit exception is made) and hence we can safely ignore this remainder. Likewise, the first term in (5.26) is due to compressibility and mean layer depth changes (via (5.19)), and we will assume that such compressible changes are relatively small, i.e. we assume that the second term in (5.26) is much bigger than the first. So, for practical purposes we approximate the impulse evolution by

$$\frac{d\mathcal{I}}{dt} = +\int (\nabla\overline{u}^L)\cdot\mathbf{p}\,dxdy. \qquad (5.28)$$

If the source term can be written as a time derivative of another quantity, then this would yield a conservation law. This is as far as we can go using the general exact GLM equations. Significantly more progress is possible if we turn to the ray tracing equations, which describe the evolution of a slowly varying wavetrain.

5.2.4 Ray Tracing Equations

We now assume that the disturbance field consists of a slowly varying wavetrain containing small-amplitude waves. This involves two small parameters, namely the wave amplitude $a \ll 1$ and another parameter $\epsilon \ll 1$ that measures the scale separation between the rapidly varying phase of the waves and the slowly varying mean flow, wavetrain amplitude, central wavenumber, and so on. The asymptotic equations that describe the leading-order behaviour of the wavetrain are the standard ray tracing equations for linear waves. We will not carry out explicit expansions in a or ϵ here because these results are well known (e.g. [13]), so we just note the outcome.

In a slowly varying wavetrain the solution looks everywhere like a plane wave locally, but the amplitude, wavenumber, and frequency of the plane wave are varying slowly in space and time. More specifically, if the fields in a wavetrain are proportional to $\exp(i\theta)$ for some wave phase θ, then the local wavenumber vector and frequency are defined by

$$\boldsymbol{k}(\boldsymbol{x}, t) = +\nabla\theta \quad \text{and} \quad \omega(\boldsymbol{x}, t) = -\theta_t. \tag{5.29}$$

Note that (5.29) implies

$$\nabla \times \boldsymbol{k} = 0 \quad \Leftrightarrow \quad \nabla \boldsymbol{k} = (\nabla \boldsymbol{k})^T, \tag{5.30}$$

which is a non-trivial statement in more than one dimension. The key asymptotic result in ray tracing is that the dispersion relation must be satisfied locally, i.e., $\boldsymbol{k}$ and ω must satisfy the dispersion relation for plane waves using the local values for the basic state. For example, the shallow-water dispersion relation for plane gravity waves with $H =$ constant and $\boldsymbol{k} = (k, l)$ is

$$\omega = \Omega(\boldsymbol{k}) = \boldsymbol{U} \cdot \boldsymbol{k} \pm \sqrt{gH}\,\kappa, \tag{5.31}$$

where $\kappa = |\boldsymbol{k}|$ is the wavenumber magnitude and $\boldsymbol{U}$ is the velocity of a constant basic flow. The basic flow induces the Doppler-shifting term $\boldsymbol{U} \cdot \boldsymbol{k}$, so the absolute frequency ω differs from the intrinsic frequency $\hat{\omega} = \omega - \boldsymbol{U} \cdot \boldsymbol{k}$. It is the intrinsic frequency that is relevant for the local fluid dynamics relative to the basic flow. In ray tracing only a single branch for the intrinsic frequency is considered in a given wavetrain; we pick the upper sign without loss of generality.

Now, if the still water depth $H(\boldsymbol{x})$ and basic flow $\boldsymbol{U}(\boldsymbol{x})$ are slowly varying[6], then (5.31) applies locally, i.e. we have

$$\omega = \Omega(\boldsymbol{k}, \boldsymbol{x}) = \boldsymbol{U}(\boldsymbol{x}) \cdot \boldsymbol{k} + \sqrt{gH(\boldsymbol{x})}\,\kappa, \tag{5.32}$$

where $\boldsymbol{k}$ and ω are defined by (5.29). Indeed, substituting (5.29) in (5.32) yields a first-order nonlinear PDE for the wave phase:

$$\theta_t + \Omega(\nabla\theta, \boldsymbol{x}) = 0 \quad \Rightarrow \quad \theta_t + \boldsymbol{U} \cdot \nabla\theta + \sqrt{gH(\boldsymbol{x})}\,|\nabla\theta| = 0. \tag{5.33}$$

---

[6] We assume that $\boldsymbol{U}(\boldsymbol{x})$ and $H(\boldsymbol{x})$ satisfy the steady nonlinear shallow-water equations.

This is the Hamilton–Jacobi equation for the wave phase. The solution of this first-order PDE involves finding the characteristics, which are the group-velocity rays along which k can be found by integrating a set of ODEs. Using the standard procedure for the characteristic system we obtain the Hamiltonian system of ODEs

$$\frac{d\boldsymbol{x}}{dt} = \boldsymbol{c}_g = +\frac{\partial \Omega}{\partial \boldsymbol{k}} \quad \text{and} \quad \frac{d\boldsymbol{k}}{dt} = -\frac{\partial \Omega}{\partial \boldsymbol{x}}, \tag{5.34}$$

where $\boldsymbol{c}_g$ is the absolute group velocity, d/dt is the rate of change along a ray, and the partial derivatives of $\Omega(\boldsymbol{k}, \boldsymbol{x})$ act on the explicit dependence of the frequency function Ω, which plays the role of the Hamiltonian function in this ODE set. The evolution of $\boldsymbol{x}$ and $\boldsymbol{k}$ describes the propagation and the refraction of the wavetrain, respectively. It is not necessary to compute θ explicitly in this procedure, although its value along a ray could be found from integrating

$$\frac{d\theta}{dt} = \boldsymbol{k} \cdot \frac{\partial \Omega}{\partial \boldsymbol{k}} - \Omega. \tag{5.35}$$

For steady U and H the Hamiltonian function $\Omega(\boldsymbol{k}, \boldsymbol{x})$ has no explicit time dependence and then it is a generic consequence of the Hamiltonian system (5.34) that $d\omega/dt = 0$, i.e. the absolute frequency $\omega = \Omega$ is constant along a ray. Of course, this does not imply that the intrinsic frequency $\hat{\omega} = \omega - \boldsymbol{U} \cdot \boldsymbol{k}$ is constant along a ray as well; indeed, the changes in $\hat{\omega}$ when $\boldsymbol{U}$ is non-uniform are crucial to the wave dynamics of refraction, for critical layers, and so on. If the basic state is slowly evolving in time as well as in space, then we have the more general $d\omega/dt = \partial\Omega/\partial t$.

   In the shallow water case the ray tracing equations come out as

$$\frac{d\boldsymbol{x}}{dt} = \boldsymbol{U} + \sqrt{gH}\,\frac{\boldsymbol{k}}{\kappa} \quad \text{and} \quad \frac{d\boldsymbol{k}}{dt} = -(\nabla\boldsymbol{U}) \cdot \boldsymbol{k} - \sqrt{g}\,\kappa\,\nabla\sqrt{H}. \tag{5.36}$$

The second, depth-related refraction term shows how components of $\boldsymbol{k}$ can be changed in the presence of a gradient in still water depth H. This is relevant for waves propagating on a beach, for instance. Note that the first, velocity-related refraction term involves a similar operator as in (5.27), i.e. the $\boldsymbol{k}$ contracts with $\boldsymbol{U}$ and not with ∇. This will turn out to be a crucial observation for the impulse budget. Incidentally, the phase evolution along a ray from (5.35) is $d\theta/dt = 0$, which is typical for non-dispersive waves.

   The ray tracing equations are completed by an equation for the wave amplitude, which in the most ideal case of a basic flow that varies slowly in all directions and in time is given by the conservation law for wave action along non-intersecting rays (i.e. away from caustics[7]):

---

[7] As is well known, at caustics neighboring rays intersect and (5.37) and the other ray tracing equations become invalid and must be replaced by more accurate asymptotic approximations; we will not consider caustics here, but see [13].

$$\frac{\partial}{\partial t}\left(\frac{\bar{E}}{\hat{\omega}}\right) + \nabla \cdot \left(\frac{\bar{E}}{\hat{\omega}}c_g\right) = 0 \quad \Leftrightarrow \quad \frac{d}{dt}\left(\frac{\bar{E}}{\hat{\omega}}\right) + \frac{\bar{E}}{\hat{\omega}}\nabla \cdot c_g = 0. \tag{5.37}$$

Here $\bar{E}$ is the phase-averaged wave energy per unit area of the waves. For example, in the shallow water case

$$\bar{E} = \frac{1}{2}\left(H\overline{u'^2} + H\overline{v'^2} + g\overline{h'^2}\right) = H\overline{|u'|^2} = g\overline{h'^2} \tag{5.38}$$

in terms of the linear wave velocity $u' = (u', v')$ and depth disturbance h'; this also shows the energy equipartition. Note carefully that the intrinsic frequency $\hat{\omega}$ appears in the definition of the wave action, not the absolute frequency ω. Because the wave field looks locally like a plane wave, knowledge of $\bar{E}$ and k implies knowledge of the amplitudes of u' and h'. Specifically, in a plane wave with $\hat{\omega} > 0$ the so-called polarization relations $u'/\sqrt{gH} = (h'/H)k/\kappa$ hold, which complete the wavetrain description.

The pseudomomentum vector takes a particularly simple form in ray tracing: all we need to do is to evaluate the GLM pseudomomentum definition $\mathsf{p}_i = -\overline{\xi_{j,i}u_j(x + \xi, t)}$ at leading order for a plane wave. For small wave amplitude, $\xi = O(a)$, and therefore the leading-order non-vanishing contribution to $\mathbf{p}$ arises at $O(a^2)$ and involves the $O(a)$ part of $u_j(x + \xi, t)$. This illustrates that $\mathbf{p}$ is a wave property, i.e. it is $O(a^2)$ for small-amplitude waves, but it can be evaluated using just the linear, $O(a)$ solution.

We write down the leading-order approximation for $\mathbf{p}$ using $u = U + u' + O(a^2)$ where $u' = O(a)$ is the linear wave velocity. Taylor-expanding $U(x + \xi, t)$ with one term yields

$$\mathsf{p}_i = -\overline{\xi_{j,i}\xi_m}U_{j,m} - \overline{\xi_{j,i}u'_j} + O(a^3). \tag{5.39}$$

So far we have used $a \ll 1$ but not $\epsilon \ll 1$. Invoking the second small parameter now allows us to neglect the first term against the second, because the gradient of U involves a small factor ϵ. Of course, this is also consistent with the idea of a local plane wave. Furthermore, in a plane wave with constant U the particle displacement evolution in (5.4) is approximated to $O(a)$ by

$$\xi_t + U \cdot \nabla \xi = u'. \tag{5.40}$$

In a plane wave this relation becomes

$$-i(\omega - U \cdot k)\xi = -i\hat{\omega}\xi = u' \tag{5.41}$$

and therefore

$$\nabla \xi = -\frac{k}{\hat{\omega}}u' \quad \Leftrightarrow \quad \xi_{j,i} = -\frac{k_i}{\hat{\omega}}u'_j, \tag{5.42}$$

where ∇ corresponds to k. Substituting back into (5.39) yields

$$\mathbf{p} = +\frac{k}{\hat{\omega}}\overline{|u'|^2}. \qquad (5.43)$$

Now, shallow-water plane waves satisfy energy equipartition by (5.38) and this leads to the final expression for the leading-order pseudomomentum in a plane wave:

$$\mathbf{p} = +\frac{k}{\hat{\omega}}\frac{\bar{E}}{H}. \qquad (5.44)$$

Thus pseudomomentum equals wavenumber vector times wave action density.[8] This shows that in a plane wave the pseudomomentum vector is always parallel to the intrinsic phase speed $\hat{c}_g = k\hat{\omega}/\kappa^2$, regardless of the sign of $\hat{\omega}$.

Actually, (5.44) is a generic relation, i.e. it applies to plane waves in all wave systems, including internal waves, internal waves with Coriolis forces, or Rossby waves (e.g. [13]). This fact is disguised in our derivation, which uses equipartition and other results that may or may not work in a given wave system. Nevertheless, the appropriate pseudomomentum definition is always such that (5.44) holds for a plane wave. This is important.

It turns out that in shallow water the leading-order pseudomomentum density for a slowly varying wavetrain is also equal to two other familiar wave properties, the so-called bolus velocity $\overline{h'u'}/H$ and the Stokes drift $\overline{u_i}^S = \overline{\xi_j u'_{i,j}}$. The derivation involves neglecting derivatives of mean fields and using the linear relation $h' = -H\nabla \cdot \boldsymbol{\xi}$ as well as $\nabla \times u' = 0$. Unlike the generic expression (5.44) above, these additional equalities are specific to the shallow-water system and do not carry over to other systems. For example, in the Boussinesq system the Stokes drift in a plane wave is zero, but not the pseudomomentum.

The ray tracing evolution law for $\mathbf{p}$ follows from multiplying (5.37) by k and using (5.34) and (5.36). The result is

$$\frac{\partial \mathbf{p}}{\partial t} + \frac{1}{H}\nabla \cdot \left(H\mathbf{p}c_g\right) = -\frac{\bar{E}}{H\hat{\omega}}\frac{\partial \Omega}{\partial x} = -(\nabla U) \cdot \mathbf{p} - |\mathbf{p}|\sqrt{g}\,\nabla\sqrt{H}. \qquad (5.45)$$

This shows that $\mathbf{p}$ inherits the conservation properties of k, in an integral sense. For example, if Ω has no explicit dependence on x, then k is constant along rays and $H\mathbf{p}_1$ satisfies an integral conservation law. Of course, whether or not a particular component of $\mathbf{p}$ is conserved in an integral sense does not affect the importance of *all* components of $\mathbf{p}$ in the GLM circulation theorem!

---

[8] The non-essential depth factor H could be absorbed in the definition of $\bar{E}$ or of $\mathbf{p}$; it arises because $\bar{E}$ is a density per unit area, which is convenient in the wave action law, whereas $\mathbf{p}$ is a density per unit mass, which is convenient in the GLM circulation theorem. The world is made imperfect.

5.2.5 Impulse Plus Pseudomomentum Conservation Law

So far we allowed the still water depth H to be variable, which is natural for shallow-water waves on beaches and other applications. However, to make progress now we restrict to constant H; perhaps at a later stage the present theory can be extended to cover variable H. (An extension to the intermediate case of one-dimensional $H(x)$ is given in Sect. 5.4.1 on wave-driven vortex dynamics on beaches.)

For constant H the pseudomomentum law (5.45) simplifies to

$$\frac{\partial \mathbf{p}}{\partial t} + \nabla \cdot \left(\mathbf{p}c_g\right) = -(\nabla U) \cdot \mathbf{p}, \tag{5.46}$$

which shows that refraction by the mean flow is now the only mechanism to change the pseudomomentum. The total pseudomomentum in the domain is

$$\mathcal{P} = \int \mathbf{p}\,dxdy, \tag{5.47}$$

and for compact $\mathbf{p}$ its rate of change is obviously

$$\frac{d\mathcal{P}}{dt} = -\int (\nabla U) \cdot \mathbf{p}\,dxdy. \tag{5.48}$$

The Lagrangian mean flow $\overline{u}^L = U + O(a^2)$, so the leading-order expression for the impulse law (5.28) is

$$\frac{d\mathcal{I}}{dt} = +\int (\nabla U) \cdot \mathbf{p}\,dxdy \tag{5.49}$$

and therefore we have the conservation law [17]

$$\mathcal{I} + \mathcal{P} = \text{constant.} \tag{5.50}$$

This conservation law is remarkable both in its simplicity and its scope, because it includes arbitrary refraction by the mean flow. The impulse $\mathcal{I}$ is a simple linear functional of the mean PV $\overline{q}^L$, so changes in $\mathcal{I}$ can be easily monitored and visualized. In general, (5.50) quantifies that net changes in pseudomomentum are compensated for by net changes in mean flow impulse. This is reminiscent of similar-sounding results for zonal momentum in zonally symmetric mean flow theory, so (5.50) extends those results to local wave–mean interaction theory. As is natural from a fluid-dynamical point of view, the zonally symmetric results are based on momentum whereas the local results here are based on vorticity.

The theoretical considerations are completed by including the effects due to a body force F, of dissipative origin or otherwise, in the momentum equation (5.1). We first define the useful disturbance-associated mean force

$$\boldsymbol{\mathcal{F}} = -\overline{(\nabla\boldsymbol{\xi})\cdot\boldsymbol{F}(\boldsymbol{x}+\boldsymbol{\xi},t)} \quad \Leftrightarrow \quad \mathcal{F}_i = -\overline{\xi_{j,i}F_j(\boldsymbol{x}+\boldsymbol{\xi},t)}. \qquad (5.51)$$

The relation between $\boldsymbol{F}$ and $\boldsymbol{\mathcal{F}}$ is analogous to the relation between $\boldsymbol{u}$ and $\mathbf{p}$. It is then possible to show that [15, 11]

$$\overline{D}^L \overline{q}^L = \frac{\nabla \times (\overline{\boldsymbol{F}}^L - \boldsymbol{\mathcal{F}})}{\tilde{h}} \qquad (5.52)$$

holds exactly and therefore the exact impulse law becomes

$$\frac{d\boldsymbol{\mathcal{I}}}{dt} = + \int (\nabla\overline{\boldsymbol{u}}^L)\cdot\mathbf{p}\,dxdy + \int (\overline{\boldsymbol{F}}^L - \boldsymbol{\mathcal{F}})\,dxdy, \qquad (5.53)$$

if $\overline{\boldsymbol{F}}^L - \boldsymbol{\mathcal{F}}$ is compact so that (5.24) can be used. Now, the reason to introduce $\boldsymbol{\mathcal{F}}$ is because this vector appears on the right-hand side of the pseudomomentum law. In particular, the ray tracing equation (5.46) with forcing is

$$\frac{\partial\mathbf{p}}{\partial t} + \nabla\cdot(\mathbf{p}\boldsymbol{c}_g) = -(\nabla U)\cdot\mathbf{p} + \boldsymbol{\mathcal{F}} \qquad (5.54)$$

and therefore

$$\frac{d\boldsymbol{\mathcal{P}}}{dt} = -\int (\nabla U)\cdot\mathbf{p}\,dxdy + \int \boldsymbol{\mathcal{F}}\,dxdy. \qquad (5.55)$$

Thus, the conservation law (5.50) is replaced by

$$\frac{d}{dt}(\boldsymbol{\mathcal{I}}+\boldsymbol{\mathcal{P}}) = \int \overline{\boldsymbol{F}}^L\,dxdy, \qquad (5.56)$$

because the terms in $\boldsymbol{\mathcal{F}}$ cancel. This shows that internal redistributions of momentum via $\boldsymbol{\mathcal{F}}$ do not affect the sum of impulse plus pseudomomentum.

The effect of viscous stresses will be considered in the next section, but for completeness we note here the results for the case in which $\boldsymbol{F}$ is due to an irrotational wavemaker, i.e. $\boldsymbol{F} = \nabla\phi$ for some compact potential ϕ. This case is particularly important for numerical experiments and for simple models of waves generated by oscillating boundaries. We have $\nabla \times \boldsymbol{F} = 0$ and therefore q remains a material invariant. Consistent with this we have the exact relations

$$\overline{\boldsymbol{F}}^L - \boldsymbol{\mathcal{F}} = \nabla\overline{\phi}^L \quad \Rightarrow \quad \overline{D}^L \overline{q}^L = 0, \qquad (5.57)$$

and therefore the mean impulse is also not explicitly affected by an irrotational wavemaker. In (5.56) the contribution of $\nabla\overline{\phi}^L$ integrates to zero for compact ϕ and so we have

$$F = \nabla\phi \quad \Rightarrow \quad \frac{d}{dt}(\mathcal{I} + \mathcal{P}) = \int \overline{F}^L \, dx dy = \int \mathcal{F} \, dx dy. \qquad (5.58)$$

This also shows that an irrotational wavemaker creates "normal" momentum and pseudomomentum at the same rate. This is a typical result also for waves generated by flow past undulating boundaries. Finally, $\overline{F}^L \approx \mathcal{F}$ for slowly varying irrotational wavemakers, because then $\nabla\overline{\phi}^L$ in (5.57a) has an explicit small factor $O(\epsilon)$.

5.3 PV Generation by Wave Breaking and Dissipation

The generation of potential vorticity by wave breaking is a very direct wave–vortex interaction: dissipating waves robustly create PV structures out of nothing. If the dissipation persists, then these PV structures can grow in time and therefore we have a strong interaction. As we shall see, in the case of a wavetrain the new PV structure resembles a vortex couple, i.e. the PV change integrates to zero and the impulse of the new PV structure is equal to the amount of pseudomomentum that has been dissipated [36, 11]. This robust result underlies the standard theory of wave drag as well, although in the standard theory the mean flow is zonally symmetric and the role of vorticity is implicit.

We consider the fluid-dynamical link between PV generation and wave breaking first, then we look at an idealized example of a wavepacket life cycle, and finally we consider the example of wave-driven longshore currents on beaches.

5.3.1 Breaking Waves and Vorticity Generation

It is a basic fluid-dynamical fact that breaking waves can generate vorticity even if there has been no vorticity prior to the breaking. A classical and vivid example is the spectacular breaking of surface waves in surfers' paradise movies. For instance, consider a two-dimensional surface wave propagating from left to right in the xz-plane and assume that the wave is steepening and overturning, say because the water depth is decreasing in x as it would be in the approach to a beach. The water–air interface forms a plunging breaker until the moment the overturning wave crest crashes onto the water just before the crest. The flow can remain essentially irrotational up to this moment. Thereafter, there are violent viscous boundary layer effects at the overturned water–water interface and a rapid transition to three-dimensional turbulence occurs, which is clearly visible and audible by the foam and bubbles in the breaking region (Fig. 5.3).

The presence of the three-dimensional turbulence alone indicates a significant creation of vorticity, but to us the more or less disorganized vorticity of the turbulence is not of particular interest. Rather, there is also an organized, large-scale component of vorticity in the y-direction, which results from the conversion of the

Fig. 5.3 Plunging wave breaker. The wave is moving *left* to *right* and the rolling motion after breaking implies an organized bundle of vorticity pointing into the picture along the wave crest, spinning clockwise

overturning irrotational water mass into a rolling water mass with an appreciable amount of solid-body rotation in the y-direction, i.e. clockwise in the xz-plane.

Now, in the strict two-dimensional version of this problem (which is physically impossible, of course, because of the three-dimensional turbulence), this organized vorticity points strictly into the y-direction. However, we can imagine that the breaking wave is confined in the y-direction such that wave breaking is possible only in $y \in [0, D]$ with some lateral breaking width D. This would be appropriate for the localized wave breaking on a beach with a point break, for instance. In this case the vorticity must be confined to this lateral interval and because vorticity lines cannot end in the fluid that means the clockwise vorticity in the core of the breaker must burrow out and connect to the boundaries of the fluid. If the water is deep enough it stands to reason that the only possible boundary is the water–air boundary. This leads to the image of a banana-shaped region of organized vorticity, with the core of the banana lying in $y \in [0, D]$ and the upwards-bending ends of the banana connecting to the free surface.

A modicum of doodling then shows that viewed from above (i.e. looking down onto the xy-plane) the vertical vorticity components are such that there is positive vorticity to the left of the propagation direction of the breaking wave and negative vorticity to the right [37, 38].[9] Consequently, the impulse associated with this vertical vorticity points in the x-direction, i.e. it points in the same direction as the phase velocity of the wave. This is consistent with a breaking-induced conversion of horizontal pseudomomentum (which is parallel to the intrinsic phase speed) into mean flow impulse.

---

[9] To understand the signs you may find it useful to consider a rolling banana at the moment when its curved ends are pointing upward.

Fundamentally the same considerations can be applied to the overturning and breaking of internal waves in a continuously stratified fluid [36], with stratification surfaces of constant entropy replacing the water–air interface and with Rossby–Ertel PV replacing the vertical vorticity. Again, the lateral confinement of the breaking region sets the width of the resultant vortex couple that is being produced on the stratification surface under consideration.[10]

5.3.2 Momentum-Conserving Dissipative Forces

Now, in order to verify the quantitative details of the exchange between pseudo-momentum and impulse during wave dissipation, we need to consider the details of $\mathcal{F}$ and $\overline{F}^L$ during wave breaking, which is hard in theory if three-dimensional turbulence is present during the breaking. A constructive middle ground is offered by considering wave breaking through shock formation in compressible fluids. For instance, it is well known in engineering gas dynamics that the flow through a curved shock, or through a shock whose strength varies along the shock front, leads to the generation of vorticity in the flow behind the shock (this is sometimes called Crocco's law in gas dynamics). In particular, if the flow is two-dimensional, then the generated vorticity is normal to the plane of the flow.

The case of a variable-strength shock front appears similar to the case of a finite-width plunging breaker, at least when viewed at some distance from above! Moreover, the shallow-water equations also describe the two-dimensional flow of a constant-entropy ideal gas with ratio of specific heats equal to 2 (see [10] for some peculiar uses of this gas-dynamical analogy), although the precise analogy breaks down at a shock because the local production of entropy at the shock cannot be captured in the shallow-water equations, which dissipates mechanical energy at the shock instead.

Nevertheless, the jump conditions at a shallow-water shock are the same as the classical jump conditions at a hydraulic bore, which is another reason to use the shallow-water equations for flows with shocks, even though the assumptions that underly shallow-water theory clearly break down at a shock. Thus we take the view that shocks in shallow water provide a reasonable model in which to study wave–vortex interactions due to wave breaking and dissipation [11]. Details are found in this reference, which also includes Coriolis forces (see also [7] for a related study on the equatorial beta-plane), so here we will just note the main results.

---

[10] By their mathematical construction both vorticity and potential vorticity always satisfy conservation laws in the integral sense, i.e. their evolution law can always be written in flux divergence form, even in the presence of external forces [24, 36]. For instance, the Rossby–Ertel PV is defined by $q = \nabla \times \boldsymbol{u} \cdot \nabla \theta / \rho$, where θ is potential temperature for atmospheric applications. This can be rewritten as $\rho q = \nabla \cdot (\theta \nabla \times \boldsymbol{u})$ for *arbitrary* choices of the scalar θ. This form makes obvious that $(\rho q)_t$ is always the divergence of some flux, even for non-physical choices of θ. In some cases this mathematical fact can be used to some advantage, but I find it mostly confusing.

Now, the jump conditions at a shock in shallow water follow from the conservation laws for mass and momentum. These conservation laws remain valid if momentum-conserving viscous stresses are added to the equations, such as would arise in the two-dimensional compressible Navier–Stokes equations, for instance. The presence of such viscous terms allows the flow fields to remain smooth at the shock, so the mathematical formalism of GLM theory can apply with $\boldsymbol{F}$ being the force due to these viscous stresses. Because $\boldsymbol{F}$ is momentum conserving in (5.1), it must be of the form

$$F_i = \frac{1}{h}\tau_{ij,j}, \tag{5.59}$$

where τ_{ij} are the components of a symmetric stress tensor. The density factor $1/h$ is crucial in this expression. Now, it can be shown that $\overline{\boldsymbol{F}}^L$ inherits this divergence form, i.e. it turns out that

$$\overline{F}_i^L = \frac{1}{\tilde{h}}\tilde{\tau}_{ij,j}, \tag{5.60}$$

where $\tilde{\tau}_{ij}$ are the components of a certain mean tensor related to τ_{ij} and the displacement fields. This is an exact result in GLM theory (e.g. [13]). The details of $\tilde{\tau}_{ij}$ are not important here, but it is important that $\overline{\boldsymbol{F}}^L$ is always proportional to the derivative of a mean field. For a slowly varying wavetrain, such a derivative brings in a small factor $O(\epsilon)$. On the other hand, $\mathcal{F}$ does not have this divergence form, so in a slowly varying regime it does not carry a small factor $O(\epsilon)$. The upshot is the generic result that

$$|\mathcal{F}| \gg |\overline{\boldsymbol{F}}^L| \tag{5.61}$$

for momentum-conserving forces within a slowly varying wavetrain. We can therefore neglect $\overline{\boldsymbol{F}}^L$ in this case, which is in contrast to the previously discussed situation for forces due to an irrotational wavemaker, in which $\mathcal{F}$ and $\overline{\boldsymbol{F}}^L$ are of the same size. This remarkably simple result implies that the impulse plus pseudomomentum conservation law holds in the presence of momentum-conserving wave dissipation, i.e. $\mathcal{I} + \mathcal{P}$ is constant even if waves dissipate.

5.3.3 A Wavepacket Life Cycle Experiment

We illustrate the results obtained so far by considering a simple life cycle of a wavepacket in shallow water. A wavepacket is a special case of a wavetrain, in which the wavenumber is essentially constant across a compact amplitude envelope; this is a single bullet of wave activity. In a wavepacket the amplitude varies more rapidly than the wavenumber vector and it has been known for a long time [8] that the wavepacket scaling is less robust than the wavetrain scaling, i.e. a wavepacket will quickly evolve into a wavetrain such that wavenumber and amplitude again vary on the same scale. Nevertheless, wavepackets are convenient conceptual building blocks and they are helpful to understand both waves and wave–vortex interactions.

Now, the life cycle begins with a quiescent fluid in which at time $t = 0$ and location $\boldsymbol{x} = (0, 0)$, say, a wavepacket is generated by an irrotational body force. The wavepacket then travels to a new location where at time $t = T$ it is dissipated by a momentum-conserving body force. Let the large wavenumber be $\boldsymbol{k} = (k, 0)$ with $k > 0$ and we use the convention $\hat{\omega} > 0$ as before. Then the pseudomomentum $\mathbf{p} = (\mathsf{p}_1, 0)$ with $\mathsf{p}_1 > 0$ and the x-component of the ray tracing law (5.54) is

$$\frac{\partial \mathsf{p}_1}{\partial t} + c \frac{\partial \mathsf{p}_1}{\partial x} = \mathcal{F}_1. \tag{5.62}$$

This uses $c = \sqrt{gH} =$ constant and $U = 0$. It is easiest to envisage a process of quite rapid wave generation, i.e. $\mathcal{F}_1$ acts impulsively at $t = 0$ to produce the wavepacket in a short time interval such that the flux term is negligible during generation and therefore $\partial_t \mathsf{p}_1 = \mathcal{F}_1$. Such rapid forcing is not essential, but it gives the easiest picture. For definiteness, we write

$$\text{wave generation:} \quad \mathcal{F} = \delta(t)(f(x, y), 0), \tag{5.63}$$

where $f(x, y) \geq 0$ is a smooth envelope for the wavepacket centred around the origin; a Gaussian would do. Directly after the generation we have $\mathsf{p}_1 = f(x, y)$. The curl of this pseudomomentum field is $\nabla \times \mathbf{p} = -\partial_y f$, which is positive to the left of the wavepacket and negative to the right. Here left and right are relative to the direction of $\mathbf{p}$. This is the characteristic signature of $\nabla \times \mathbf{p}$ for a wavepacket. For definiteness, we let f integrate to unity so $\mathcal{P} = (1, 0)$ at this stage. The impulse $\mathcal{I} = 0$ because $\overline{q}^L = 0$, of course.

After generation, the wavepacket propagates without change in shape in a straight line with constant speed c from left to right, i.e., $\mathsf{p}_1 = f(x - ct, y)$. Both $\mathcal{I} = 0$ and $\mathcal{P} = (1, 0)$ remain constant during propagation. The subsequent dissipation process can be modelled either impulsively as well or by an exponential attenuation such that $\mathcal{F}_1 = -\alpha \mathsf{p}_1$, where $\alpha > 0$ is an exponential damping rate. The latter option is familiar from linear wave dissipation mechanisms. Either way, the sum $\mathcal{I} + \mathcal{P} = (1, 0)$ remains constant because the dissipation is momentum conserving. In other words, any pseudomomentum lost to dissipation is converted into impulse.

How does the mean flow react in detail to this wavepacket life cycle? We are really only interested in the vortical part of the mean flow response, but in shallow water it is easy enough to write down the complete mean flow equations at $O(a^2)$ for small-amplitude slowly varying wavetrains (e.g. [14]). In the present case they are

$$\tilde{h}_t + H\nabla \cdot \overline{\boldsymbol{u}}^L = 0 \quad \text{and} \quad \overline{\boldsymbol{u}}_t^L + g\nabla \tilde{h} = \mathbf{p}_t - \frac{1}{2}\nabla \overline{|\boldsymbol{u}'|^2} + \overline{\boldsymbol{F}}^L - \mathcal{F}. \tag{5.64}$$

A simple derivation of (5.64) uses that $\mathbf{p} \approx \overline{h'\mathbf{u}'}/H$ and $\tilde{h} \approx \overline{h}$ for slowly varying wavetrains; taking the curl of (5.64) recovers (5.52) at $O(a^2)$, in which $\overline{\mathbf{u}}^L = O(a^2)$ and $\tilde{h} = H$ to sufficient approximation. The complete set (5.64) illustrates that the mean flow inherits a forced version of the modal structure of the linear equations, i.e. there are two gravity-wave modes and one balanced, vortical mode. It also shows that the mean flow forcing can be viewed as being due to a combination of transience, wavetrain inhomogeneity, external forcing, and dissipation.

During the rapid generation of the wavepacket the third and fourth forcing terms cancel and the first term dominates the second term because of the time derivative. The same argument applies to the left-hand side of (5.64) and therefore we have $\tilde{h} = H$ and $\overline{\mathbf{u}}^L = \mathbf{p} = (f, 0)$ just after the impulsive wavepacket generation. This initial condition for the mean flow is a compact bullet of x-momentum centred at the origin.

The subsequent evolution of the mean flow consists of the evolution of this initial condition under the additional influence of the forcing terms. The upshot is that there is a persistent generation of weak $O(a^2)$ mean flow gravity waves during the propagation of the two-dimensional wavepacket [8].[11] The vorticity of the mean flow remains bound to the wavepacket because of $\overline{q}^L = 0$ and therefore $\nabla \times \overline{\mathbf{u}}^L = \nabla \times \mathbf{p}$; a mean flow vorticity probe would detect a vorticity couple flanking the wavepacket, but this vorticity couple would move with the linear wave speed c, not the nonlinear advection speed.

After $t = T$ dissipation becomes active; during dissipation $\overline{\mathbf{F}}^L$ is negligible and $\mathcal{F} = -\alpha\mathbf{p}$, say. Then (5.62) leads to $\mathbf{p}_1 = f(x - ct, y)\exp(-\alpha(t - T))$ for $t \geq T$. During this attenuation process we have

$$H\overline{q}_t^L = -\nabla \times \mathcal{F} = \alpha\nabla \times \mathbf{p}, \tag{5.65}$$

which shows how $\nabla \times \mathbf{p}$ is transferred into $\overline{q}^L$ in a manner that is consistent with the conservation of $\mathcal{I} + \mathcal{P} = (1, 0)$. The form of (5.65) indicates that $\overline{q}^L$ evolves "as if" an effective mean force equal to minus the dissipation rate of pseudomomentum were acting on the mean flow; this effective force points in the same direction as $\mathbf{p}$. After a long time $t - T \gg 1/\alpha$ the wavepacket is practically gone, and so is the wavelike part of the mean flow, which will propagate away from the dissipation site of the wavepacket. What remains behind after $\mathbf{p} \to 0$ is the vortical mean flow with vorticity $\nabla \times \overline{\mathbf{u}}^L = H\overline{q}^L$. This is a dipolar vortex couple with $\mathcal{I} = (1, 0)$ that resembles a smeared-out version of the instantaneous curl of the wavepacket's pseudomomentum. The smearing-out is due to the advection of the wavepacket during dissipation, it can be reduced by making the dissipation more rapid.

---

[11] Actually, there is also a non-essential resonance effect because the wavepacket forcing terms move with speed c, which is the only available speed in the non-dispersive shallow-water equations. This projects resonantly onto the mean flow gravity waves, which is also a reminder of the beginning of shock formation in shallow water. Still, for our purposes this resonance is artificial and we will not consider it. For example, by adding some dispersion to the equations (e.g. by adding Coriolis forces) this resonance would disappear.

The take-home message from this wavepacket life cycle is that irrotational wave generation changes $\mathcal{P}$ but not $\mathcal{I}$, that propagation through a quiescent medium preserves both $\mathcal{P}$ and $\mathcal{I}$, and that dissipation leads to a zero-sum transference of $\mathcal{P}$ into $\mathcal{I}$. The lasting vortical mean flow response is described by $\overline{q}^L$ and behaves in an easy, generic manner. The full mean flow response also contains parts to do with wavelike mean flow dynamics, which are not generic and complicated.

If the life cycle is repeated many times then $\overline{q}^L$ would grow secularly at the dissipation site, which would in time lead to the spin-up of a substantial vortex couple there. This is a model example of a strong wave–vortex interaction.

5.3.4 Wave Dissipation Versus Mean Flow Acceleration

The life cycle experiment in the previous section makes it easy to discuss the interesting relationship between wave dissipation and mean flow acceleration (see [17] for more details). It is clear from the mean PV law (5.52) and (5.61) that there is a direct link between (momentum-conserving) wave dissipation and mean PV changes. However, does the mean flow actually accelerate where and when the waves dissipate?

We can get a clear answer to this question if we make the dissipation as impulsive as the wave generation, i.e. if we assume that the wavepacket is annihilated in a very short time interval $\Delta t \propto 1/\alpha$ such that the dominant balance in (5.62) is again $\partial_t \mathbf{p}_1 = \mathcal{F}_1$. What happens to the mean flow acceleration in this case can now be read off from (5.64): we obtain

$$\overline{u}_t^L + g\nabla\tilde{h} = -\frac{1}{2}\nabla\overline{|u'|^2}. \tag{5.66}$$

The sole remaining term on the right-hand side is bounded and vanishes during the decay interval Δt. Moreover, this term was already there prior to the wave dissipation, so it really has nothing to do with dissipation. We are led to conclude that $\overline{u}^L$ *does not change during dissipation,* i.e. there is no obvious mean flow acceleration when and where the waves are dissipating.

This is a surprising result, because it completely disagrees with the standard situation in wave drag studies in which there is a steady wavetrain subject to localized dissipation. In this case $\mathbf{p}_t = 0$ by assumption and therefore (5.66) is replaced by

$$\overline{u}_t^L + g\nabla\tilde{h} = -\frac{1}{2}\nabla\overline{|u'|^2} - \mathcal{F} = -\frac{1}{2}\nabla\overline{|u'|^2} + \alpha\mathbf{p} \tag{5.67}$$

in the dissipation region. The first term can be balanced by a suitable depth variation $\tilde{h}$, but the second term is not irrotational in general and therefore it keeps on pushing. In this case, the site of wave dissipation is also the site of mean flow acceleration, or at least of mean flow forcing.

So how can these two results be consistent with each other? They must be consistent because we can approximate a steady wavetrain by sending in an infinite sequence of wavepackets, like pearls on a string. The answer lies in the physical nature of mean flow forcing by the waves. At a fundamental level, the mean flow responds to wave-induced fluxes of mass and momentum in the shallow-water equations. Typically, these wave-induced fluxes are nonzero but also non-divergent within a plane wave. Therefore, as a wavepacket arrives at a given location, there is a period of transition from zero flux to constant flux inside the bulk of the wavepacket [34]. It is during this transient period that the mean flow acceleration is strongest in response to the nonzero flux divergences; this effect is described by $\mathbf{p}_t$ in (5.64).

Once the wavepacket leaves there is again a transient period, but this time all the flux divergences have their signs reversed, which undoes the previous mean flow accelerations. The net result is that the mean flow has been nudged back and forth by the arrival and departure of the wavepacket, which may set off some mean flow waves, but there has been no obvious lasting change.

How does dissipation change this picture? Dissipation weakens the wavepacket without leading by itself to a nonzero flux divergence. In other words, dissipation reduces the wave-induced fluxes inside the wavepacket, but it does not change their spatial distribution. So, as the wavepacket leaves the location under consideration, the undoing of the mean flow accelerations that occurred during the wavepacket's arrival are only partially undone, because by now the wavepacket has been weakened. This leads to a net residuum of the time-integrated mean flow forcing and therefore to a net lasting change in the mean flow. It is also clear that the net residual forcing points in the direction of $\mathbf{p}$.

So there is no intrinsic link between wave dissipation and mean flow forcing but there is such a link between wave transience and mean flow forcing. Mathematically, this state of affairs can perhaps most easily be spotted in the evolution law for $\nabla \times \overline{\boldsymbol{u}}^L$, which is

$$\frac{\partial}{\partial t} \nabla \times \overline{\boldsymbol{u}}^L = \frac{\partial}{\partial t} \nabla \times (\mathbf{p} - \mathcal{F}). \qquad (5.68)$$

Without dissipation $\nabla \times \overline{\boldsymbol{u}}^L$ is slaved to the pseudomomentum curl, and therefore it changes directly through wave transience. During rapid dissipation, on the other hand, the terms on the right-hand side cancel and $\nabla \times \overline{\boldsymbol{u}}^L$ does not change at all.

The apparently different result for a steady wavetrain is then explained by the necessary coalescence of wave dissipation regions and wavepacket transience regions in the case of a steady wavetrain. For instance, in the pearls-on-a-string approach many wavepackets mimic a steady wavetrain and then the fluid in the dissipation site is repeatedly accelerated by arriving wavepackets, but it never gets decelerated because the wavepackets never leave, because of the dissipation. This leads to the persistent, growing mean flow acceleration in the dissipation site that is familiar from the standard steady wavetrain picture.

If we want to summarize the main point of this section in one sentence, then it could be that wave dissipation makes irreversible a mean flow acceleration due to wave transience that has already taken place. To model this aspect of wave–vortex interaction correctly is important in order to understand the full mean flow response, including mean flow waves, during wavepacket dissipation (e.g. [44, 41, 17]).

5.4 Wave-Driven Vortices on Beaches

The material in this section is based on [14] and [5]. The breaking of ocean waves that are obliquely incident on a beach can drive longshore currents along the beach. These currents are of appreciable magnitude, with typical alongshore speeds of 1 m/s and typical horizontal current width of a hundred metres or so. Longshore currents can interact with and co-produce rip currents, they contribute to beach erosion and evolution, and they can be important in their impact on engineering structures in the nearshore region.

The basic physical mechanism for longshore currents is the wave-induced transport of alongshore momentum towards the beach and the associated wave drag when the waves are breaking in the surf zone. This momentum flux is $M = H\overline{u'v'}$ in shallow-water theory, where x is the cross-shore and y is the alongshore coordinate such that the shoreline corresponds to $x =$ constant, say (see Fig. 5.4). In the simplest geometry we allow only one-dimensional topography such that the still water depth $H(x)$ is a function only of distance to the shoreline. We also assume that the

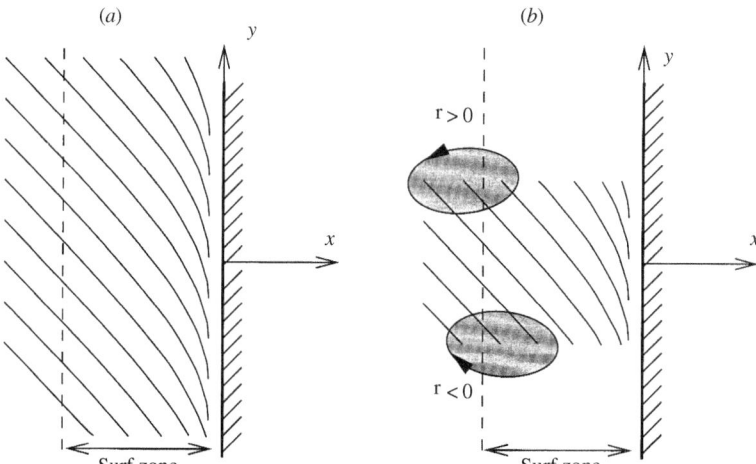

Fig. 5.4 *Left:* crests of homogeneous wavetrain obliquely incident on beach with shoreline on the *right*. The waves break in the surf zone and drive a longshore current in the positive y-direction there. There are no vortices. *Right:* crests of inhomogeneous wavetrain. The breaking location is flanked by a vortex couple generated by wave breaking; the indicated vorticity signatures are the vertical outcropping of the three-dimensional vorticity banana described in Sect. 5.3.1. Due to the oblique wave incidence, the vortex couple is slightly tilted relative to the shoreline and therefore it has a positive impulse in the y-direction

flow is periodic in the alongshore y-direction. The general case of two-dimensional topography with still water depth $H(x, y)$ is more complicated, because then there can be pressure-related momentum exchanges with the ground.

5.4.1 Impulse for One-Dimensional Topography

Can we define a useful mean flow impulse in the case of variable $H(x)$? In the case of constant H we modelled the mean flow impulse on the classical impulse for two-dimensional incompressible flow. This corresponds to a shallow-water flow between two parallel rigid plates with constant distance H. In the present case, we can look for inspiration in the case of a two-dimensional rigid-lid flow with non-uniform H. This flow is governed by

$$\nabla \cdot (H\boldsymbol{u}) = 0 \quad \text{and} \quad \frac{Dq}{Dt} = 0 \quad \text{where} \quad q = \frac{\nabla \times \boldsymbol{u}}{H}. \tag{5.69}$$

There is only a single degree of freedom in the initial-value problem, namely the vortical mode described by the initial distribution of q; the rigid lid filters all gravity waves. The corresponding two-dimensional momentum equation is

$$\frac{D\boldsymbol{u}}{Dt} + \nabla p = 0, \tag{5.70}$$

where p is the pressure at the rigid lid, which can be computed from an elliptic problem just as the pressure in incompressible flow. If we allow for $H(x)$ only, then the y-component of momentum is conserved, i.e.

$$\frac{d}{dt} \int H v\, dx dy = 0 \tag{5.71}$$

in a periodic channel geometry with solid walls at two locations in x, say. This leads to a conserved impulse in terms of the PV if we define a potential $L(x)$ for the topography such that

$$\frac{dL}{dx} = -H(x). \tag{5.72}$$

The y-component of the impulse is then (e.g. [26, 25])

$$I_2 = \int L(x) H(x) q\, dx dy \quad \Rightarrow \quad \frac{dI_2}{dt} = 0. \tag{5.73}$$

The proof uses $DL/Dt = -Hu$, integration by parts, periodicity in y, and that $u = 0$ at the channel side walls. In the constant-depth case $H = 1$, we have $L(x) = -x$ and therefore (5.73) recovers the classical impulse. On a planar constant-slope beach with $H = x$, say, we obtain $L = -x^2/2$ and so on. In general, I_2 equals the net y-momentum in (5.71) up to some constant terms related to the (constant) circulation along the channel walls.

To illustrate (5.73), we again consider I_2 due to a point vortex couple with circulations $\pm\Gamma$ and separation distance d in the x-direction.[12] Now, if the left vortex has positive circulation, then in the case of constant $H = 1$ this produces $I_2 = d\Gamma$. For variable H we obtain $I_2 = \Delta L\Gamma$ instead, where ΔL is the difference of $L(x)$ between the two vortex locations. If $H(x)$ is smooth then using the definition of $L(x)$ and the mean value theorem this can be written as $\Delta L = dH(x_*)$, where x_* is an intermediate x-position between the vortices. For small x-separations this suggests the approximation $\Delta L \approx d\tilde{H}$ where $\tilde{H}$ is the average depth at the two vortex positions and therefore $I_2 \approx d\tilde{H}\Gamma$.

The simplest example in which variable topography gives a non-trivial effect is in a domain with two large sections of constant $H = H_A$ and $H = H_B$, say, connected by a smooth transition. In this case the conservation of I_2 implies that a vortex couple that slides from one section to the other must change its separation distance. Specifically, if the couple starts in the section with $H = H_A$ and separation d_A then we have

$$I_2 = \text{const.} \quad \Rightarrow \quad d_A H_A = d_B H_B \quad \Rightarrow \quad d_B = d_A \frac{H_A}{H_B} \qquad (5.74)$$

if the couples makes it to the other section. If $H_B > H_A$, i.e. if the couple moves into deeper water, then $d_B < d_A$ and therefore the couple has moved closer together. Because the mutual advection velocity is proportional to Γ/d, this implies that the vortex couple has sped up. Considerably more detailed analytical results about the vortex trajectories can be computed in the case of a step topography [25].

Similarly, on a constant-slope beach with $H = x$ and $L = -x^2/2$, the conclusion would be that $\tilde{x}d$ is exactly constant, where $\tilde{x}$ is the average x-position of the two vortices. This has the consequence that the cross-shore separation d of a vortex couple climbing a planar beach towards the shoreline (i.e. propagating towards $x = 0$ if $H = x$) would increase as the water gets shallower.

We return to wave–vortex interactions: based on the rigid-lid role model, we define a shallow-water mean flow impulse in the y-direction at $O(a^2)$ by

$$\mathcal{I}_2 = \int L(x)H(x)\overline{q}^L \, dxdy \quad \Rightarrow \quad \frac{d}{dt}(\mathcal{I}_2 + \mathcal{P}_2) = 0 \qquad (5.75)$$

under unforced evolution or momentum-conserving dissipation. Clearly, this assumes that the mean flow behaves approximately as if there was a rigid lid, i.e. it assumes

---

[12] Strictly speaking, a point vortex model is not well posed if ∇H is nonzero, because of the infinite self-advection of a point vortex on sloping topography, which is analogous to the infinite self-advection of a curved line vortex in three dimensions. We can resolve this by replacing the point vortex with a vortex with finite radius b provided that b is much smaller than d or any other scale in the problem.

that $\nabla \cdot (H\overline{\boldsymbol{u}}^L) = 0$ and therefore mean flow gravity waves are weak. This conservation law also holds for wave refraction by the mean flow, the only difference is that the production term $-(\nabla U) \cdot \mathbf{p}$ in (5.46) and the definition of the net pseudomomentum $\mathcal{P}$ in (5.47) both acquire a factor of $H(x)$.

5.4.2 Wave-Induced Momentum Flux Convergence and Drag

Now, returning to the wave-driven currents, M is the mean wave-induced flux of y-momentum in the x-direction. It is a basic exact result in GLM theory that this "off-diagonal" mean momentum flux equals the flux of y-pseudomomentum in the x-direction [3]. Indeed, in the wavetrain regime it is easy to check that

$$M = H\overline{u'v'} = H\mathsf{p}_2 u_g = \frac{lk}{\kappa^2}\bar{E}. \tag{5.76}$$

The y-component of the pseudomomentum law (5.45) shows that

$$\partial_x(H\mathsf{p}_2 u_g) + \partial_y(H\mathsf{p}_2 v_g) = 0 \tag{5.77}$$

for a steady wavetrain. During the approach of a shoreline the wavetrain is refracted toward the beach, i.e. the wavenumber vector is turned normal to the shoreline. This implies that $|k|$ is much bigger than $|l|$ near the shoreline, which allows making a small-angle approximation in which terms l^2/k^2 and higher are neglected. In this approximation the pseudomomentum law implies $\partial_x(H\mathsf{p}_2 u_g) = 0$ and therefore M is constant for a steady wavetrain.

However, diminishing H leads to an increase in wave amplitude as measured, say, in the relative depth disturbance h'/H, which is a useful definition of non-dimensional wave amplitude. This follows from $\bar{E} = g\overline{h'^2}$, the constancy of (5.76), the ray invariance of l and $\omega = \sqrt{gH}\kappa$, and the small-angle approximation, which together imply the scaling

$$\bar{E} \propto \kappa^2/(kl) \propto \kappa \propto H^{-1/2} \quad \Rightarrow \quad h'/H \propto H^{-5/4}. \tag{5.78}$$

This indicates the sharp growth of wave amplitude as the water depth decreases, which must lead to nonlinear wave breaking before the shoreline is reached. Where the waves break the momentum flux M is diminished and therefore in the breaking region there is a net acceleration of the mean flow along the beach, which leads to the longshore current. As before, this is a momentum-conserving transfer of alongshore pseudomomentum into alongshore mean flow impulse.

The first rational theory for longshore currents was formulated by Longuet–Higgins in [32, 33]. In this theory the flow is periodic in the alongshore direction and, most importantly, the incoming wave forms a slowly varying wavetrain with constant amplitude in the alongshore direction. That means that y-derivatives of all mean quantities are zero by assumption. The mean flow situation is then described by the simplified y-component of (5.67), which is

$$\bar{v}_t^L = -\mathcal{F}_2. \qquad (5.79)$$

This very simple form stems from the assumption of a one-dimensional wavetrain. The force term $-\mathcal{F}_2$ is deduced from a saturation assumption, which limits the surface elevation h' to be less or equal to a fixed fraction of the local still water depth $H(x)$.

Actually, (5.79) describes the secular spin-up of the longshore current, but not its steady state. The forcing term is $O(a^2)$ but after a long time $t = O(a^{-2})$ the current would have grown to $O(1)$. This is a strong wave–vortex interaction, with the current playing the role of the vortices. A forced-dissipative steady state is possible if friction terms are added in (5.79). The friction can be due to both bottom friction and/or horizontal eddy diffusion by turbulent motion, but the simplest model uses bottom friction only. The usual assumption is that a turbulent boundary layer exists near the ground such that the net drag on the water column is proportional to the quadratic force $-\boldsymbol{u}|\boldsymbol{u}|$, which means that a body force $\boldsymbol{F}_B = -c_f \boldsymbol{u}|\boldsymbol{u}|/h$ must be added to the shallow-water equations. A typical value of the friction parameter is $c_f = 0.01$.

The phase average of $\boldsymbol{F}_B$ is complicated even for a plane wave on a uniform current because of the absolute value sign. It was argued in [32] that the dominant term in $\overline{\boldsymbol{F}_B}$ comes from a product of wave and mean flow contributions. This means that in order to balance the $O(a^2)$ forcing term in (5.79) the mean flow had to be $O(a)$, which yields a non-trivial scaling for the steady longshore current, i.e. the amplitude of the steady longshore current is proportional to the amplitude of the incoming wave.

5.4.3 Barred Beaches and Current Dislocation

The theory of Longuet–Higgins together with various extensions and refinements has been very successful in predicting the current structure, at least on beaches with simple topography profiles such that $H(x)$ decreases monotonically with decreasing distance to the shoreline. However, there has been evidence (e.g. [18]) that this theory is making less correct predictions on barred beaches, where there is an off-shore minimum of $H(x)$ on a topography bar crest plus a depth maximum closer to the shoreline at the bar trough (see Fig. 5.5b). Frequently, though not always, observations on such beaches show a current maximum at the location of the bar trough, i.e. at a location which is not identical with the location of strongest wave breaking. This current dislocation cannot be explained by the original theory of Longuet–Higgins.

In coastal oceanography, the presently favored explanation is the invocation of so-called wave rollers, which are meant to represent a certain body of rolling water during wave breaking that is capable of "storing" momentum for a while. The stored momentum is then released at some later time after breaking, which can be chosen so as to produce the observed current dislocation effect. As is to be expected, applying this ad hoc wave roller procedure requires a substantial amount of parameter fitting in order to produce the observed currents. Moreover, the procedure does not explain why wave rollers should be important only on barred beaches, i.e. it does not explain why the original theory, without rollers, is successful on other beaches.

Regardless of the merit and ultimate fate of wave roller models, it is also interesting to look for other mechanisms that could explain the current dislocation. One suggestion is wave-driven vortices. This goes back to comments by Peregrine, who some time ago advocated theoretically and with careful photography that more attention should be paid to vortices in nearshore dynamics [37, 38].

Vortices appear the moment one drops the crucial assumption that the wavetrain is y-independent [14, 9, 28]. Such alongshore inhomogeneity could be either through y-dependent topography or due to a y-dependent wavetrain envelope. The latter case is easier to study and was explored in [14]. Basically, the breaking of obliquely incident waves now produces a vortex couple that is oriented at a small angle to the cross-shore direction (see right panel in Fig. 5.4). The small angle of the vortex couple goes together with a small alongshore component of its impulse. This is the familiar picture discussed previously, and it constitutes a strong wave–vortex interaction just as before.

With the same model for quadratic bottom friction as before, the vortices now grow in amplitude until one of two things happens: either friction terminates their further growth or the mutual interaction between the vortices begins to move them nonlinearly. If the first alternative prevails then the flow is simply steady, and the alongshore-averaged current structure does not differ much from the predictions of Longuet–Higgins. However, if the second alternative prevails then the vortices move away from the forcing site and they take the current maximum with them (see Fig. 5.5). For a given wavetrain shape it is the size of the coefficient for bottom friction that decides which of the two alternatives is realized. It appears that for the typical $c_f = 0.01$ vortex mobility is possible, so realistic wave-driven vortices on beaches should be capable of moving around. As they move, they conserve their alongshore impulse.

Most interestingly, there are good fluid-dynamical reasons why mobile vortex couples should then "prefer" the deep water of the bar trough [14], which provides a ready explanation for current dislocation on a barred beach. In essence, a vortex couple that moves into deeper water is pushed together by the convergent horizontal motion of the water column. This reduction in distance intensifies their mutual interaction and makes them move faster and further. Conversely, a vortex couple that moves into shallower water gets more separated and slows down, which allows friction effects to take over. This also explains why on a plane beach with monotone $H(x)$ the vortices will not move far up the beach and therefore why there is little current dislocation on such a beach. This two-sided explanation is perhaps the most attractive feature of a vortex-based theory of longshore currents: *vortices like deep water.*

Finally, one could imagine that wave-driven vortices on a beach should also be capable of performing the dynamical manoeuvres commonly associated with two-dimensional turbulence. This would lead to an interesting statistical problem about

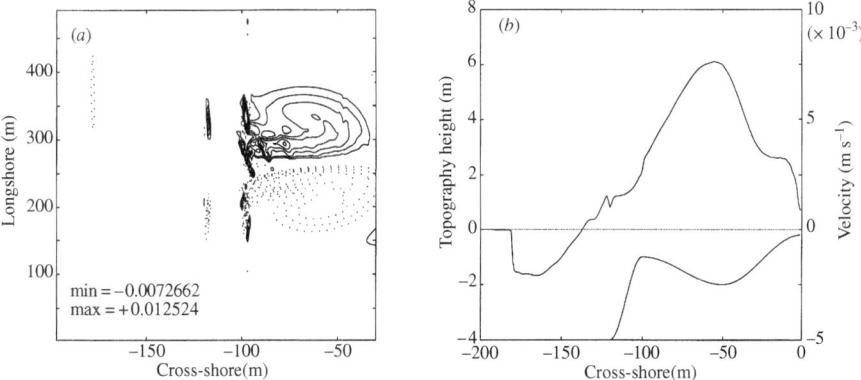

Fig. 5.5 *Left:* PV structure at a late time for inhomogeneous wavetrain incident on barred beach. The topography is indicated on the *right*. The vortex couple has moved to the bar trough, away from the main wave breaking region over the bar crest at approximately 100m off-shore. *Right:* topography and longshore current velocity profile obtained by averaging over the alongshore direction. The current maximum has been dislocated into the trough

the mechanics of the vortices in a highly non-uniform shallow-water domain. However, it turns out that $c_f = 0.01$ is too high to allow two-dimensional turbulence [5]. This follows from the investigations of two-dimensional turbulence forced at small scales and dissipated at large scales by [21]. They show that quadratic friction introduces a stopping scale in wavenumber space, i.e. it produces a barrier in wavenumber magnitude below which there can be no turbulence.

For the typical beach parameters this stopping scale more or less coincides with the wavenumber threshold below which shallow-water theory becomes accurate. In other words, vortex motions on realistic beaches that are accurately described by shallow-water models are non-turbulent. This agrees with numerical evidence, which shows non-trivial but laminar trajectories for beached vortices unless c_f is lowered substantially below 1%.

5.5 Wave Refraction by Vortices

We now turn to the exchange mechanism between mean flow impulse and pseudomomentum described by the refraction term $-(\nabla U) \cdot \mathbf{p}$ in (5.46). There is no essential need for either dissipation or topography effects here, so we can set $\mathbf{F} = 0$ and $H = 1$ for simplicity. We begin by looking in a little bit more detail at the wave refraction process and its partial analogy with passive tracer transport, then we look at refraction due to a single vortex in shallow water and the concomitant feedback on the vortex [16]. This is followed by a new system, the three-dimensional Boussinesq equations, where a different strong interaction effect is possible [27, 4, 17]. We finish by looking at the peculiar interactions between a vortex couple and a wavepacket.

5.5.1 Anatomy of Wave Refraction by the Mean Flow

Wave refraction by vortical mean flows can be easily observed in a bath tub: surface waves propagating towards a bath-tub vortex over the open plug-hole are refracted, and their tilting wave crests give a vivid, if often misleading, impression of the spinning flow around the vortex. The impression is often misleading because the wave crest pattern spins mostly clockwise if the vortex spins anti-clockwise and vice versa (see Sect. 5.5.2).

In ray tracing the refraction of the wave phase θ is described by the ray tracing equation for the wavenumber vector $k = \nabla\theta$, i.e.

$$\frac{d\boldsymbol{k}}{dt} = -(\nabla \boldsymbol{U}) \cdot \boldsymbol{k}. \tag{5.80}$$

The refraction of k is then inherited by $\mathbf{p} = k\bar{E}/\hat{\omega}$. An intuitive understanding of (5.80) can be obtained by comparing it with the more familiar evolution of the gradient $\nabla\phi$ of a passive tracer ϕ. We use the short-hand $D_t = \partial_t + U \cdot \nabla$ for the material derivative based on the basic flow U and we assume that ϕ is passively advected by U. This leads to

$$D_t\phi = 0 \quad \Rightarrow \quad D_t(\nabla\phi) = -(\nabla U) \cdot \nabla\phi. \tag{5.81}$$

Comparing (5.80) and (5.81) we find the same operator on the right-hand side. This shows that the basic flow refracts the phase θ *as if* the phase was a passive tracer relative to the basic flow U. For instance, this analogy implies that the antisymmetric part of ∇U seeks to rotate k without changing its length around an axis parallel to $\nabla \times U$. On the other hand, the symmetric part of ∇U seeks to strain the θ-surfaces, which changes both the orientation and the length of k.

This standard straining behaviour is illustrated in Fig. 5.6, which illustrates the three types of straining depending on the sign of

$$D = U_x^2 + \left(\frac{V_x + U_y}{2}\right)^2 - \left(\frac{V_x - U_y}{2}\right)^2. \tag{5.82}$$

This uses $U_x + V_y = 0$. Perhaps the most interesting message from Fig. 5.6 is that in the case of $D > 0$ the gradient of the tracer is asymptotically oriented in a direction completely determined by ∇U, i.e. the tracer gradient forgets its initial state. If the same is true for k and by implication for $\mathbf{p}$, then a wavepacket forgets its initial pseudomomentum if it is refracted by a similar flow.

Of course, the analogy between θ and ϕ is only a partial analogy for two reasons. First, the phase pattern also moves relative to the flow by the intrinsic wave propagation, so it is clearly not a passive tracer. Second, the two equations agree on the right-hand side but not on the left-hand side because (5.81) involves the $O(1)$ material derivative D_t whereas (5.80) involves the derivative along group velocity rays. These two operators differ by the advection with the intrinsic group

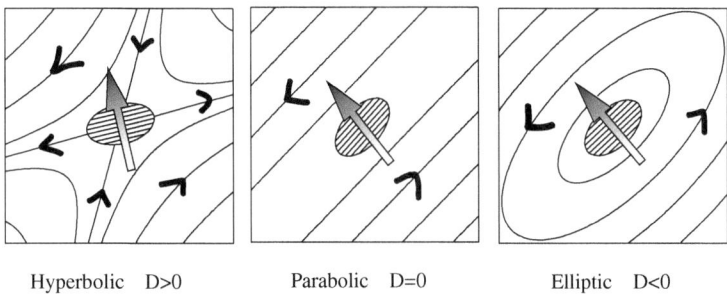

Hyperbolic   D>0   Parabolic   D=0   Elliptic   D<0

Fig. 5.6 Straining pictures depending on the sign of D in (5.82). The *streamlines* are drawn in a frame moving with the local velocity and also shown is a patch of an advected tracer ϕ together with its gradient $\nabla\phi$ indicated by an arrow. *Left: $D > 0$, hyperbolic case, open streamlines.* The tracer contours align with the axis of extension, the tracer gradient turns normal to this axis and grows exponentially in time. *Middle: $D = 0$, parabolic case* with shear flow. Tracer contours align with shear direction and $\nabla\phi$ grows linearly in time. *Right: $D < 0$, elliptic,* vorticity-dominated flow with *closed streamlines.* Tracer contours rotate in time and $\nabla\phi$ oscillates in direction and magnitude

velocity, i.e.

$$\frac{d}{dt} - \mathrm{D}_t = \widehat{\boldsymbol{c}}_g \cdot \nabla. \qquad (5.83)$$

Nevertheless, the analogy is quite useful for understanding refraction by the basic flow and it is also quantitatively relevant if the intrinsic wave propagation is weak.

5.5.2 Refraction by Weak Irrotational Basic Flow

Significant analytical progress with little effort can be made if the basic flow is irrotational, which covers the important case of flows outside of vortex cores. Simplest of all is the case of a steady weak irrotational flow, in which the Froude or Mach number $|U|/c = O(\epsilon)$ where $c = \sqrt{gH}$ and $\epsilon \ll 1$. In this case layer depth variations in the basic state are $O(\epsilon^2)$ by the usual scaling based on the Bernoulli function, and these small variations can be ignored. Furthermore, to $O(\epsilon)$ we can make use of the classical result that non-dispersive wave rays through an irrotational background flow are straight lines to $O(\epsilon)$, i.e. to first order in Froude (or Mach) number (e.g. [31], p. 261; the result readily generalizes to dispersive waves with large intrinsic group velocities relative to the background flow, see, e.g., [20]). This remarkable result (which can be derived from Fermat's principle of least time) says that whilst $\boldsymbol{k}$ and hence the intrinsic group velocity are changed by refraction due to U, the absolute group velocity $\boldsymbol{c}_g = U + c\boldsymbol{k}/\kappa$ remains pointing in the same direction.

This allows solving the ray tracing equations for $(\boldsymbol{x}, \boldsymbol{k})$ analytically to $O(\epsilon)$. The result is [16]

$$x = x_0 + \left(U_0 + c\frac{k_0}{\kappa_0}\right)s + O(\epsilon^2) \quad \text{and} \quad k = k_0 - \frac{\kappa_0}{c}(U - U_0) + O(\epsilon^2) \quad (5.84)$$

where U is evaluated along the ray, the subscript zero refers to the initial conditions at the start of the ray, and $s \geq 0$ is the distance along the straight ray. It is easy to check that k satisfies (5.80) to $O(\epsilon)$ because $U_{i,j} = U_{j,i}$.

We can apply (5.84) to the case of a single circular vortex with counterclockwise circulation $\Gamma > 0$ and radius b centred at the origin of the coordinate system. The basic flow outside the vortex is then[13]

$$U = (U, V) = \Gamma \frac{(-y, +x)}{x^2 + y^2}, \qquad (5.85)$$

and we consider the fate of a wavepacket that starts from far away (such that U_0 can be ignored) and then propagates past the vortex. For simplicity we assume that the packet does not enter the vortex core where $\nabla \times U \neq 0$. Let us denote by α the angle that the straight propagation direction of the packet makes with the location of the vortex core. For definiteness, let $\alpha > 0$, so the vortex lies on the port side of the packet. Initially, when α is very small, the refraction is turning k counterclockwise, which makes the wavepacket crests glance towards the vortex. As α reaches 45 degrees, this glancing reaches its maximum. Thereafter the crests are now turned *clockwise* until α reaches $135°$. Thereafter the crests are straightened out again and the vortex is left behind.

In the bathtub example the most visible wave refraction part is the counter-rotating crest tilt as the waves are near the vortex and $\alpha \in [45, 135]$ degrees. This counter-rotation of the wave crests gives the misleading impression of the vortex rotation sense. Of course, there are also other effects at higher order in ϵ. For example, it is shown in [16] that at $O(\epsilon^2)$ there is an irreversible scattering of the wavepacket towards the lee of the vortex.

5.5.3 Bretherton Flow and Remote Recoil

We now turn to the wave–vortex interactions that go together with refraction by a single vortex. Here we follow [16] and truncate the wavepacket trajectory by putting irrotational wave sources and sinks a finite distance L apart. Moreover, we consider a steady wavetrain generated by the source "loudspeaker" and absorbed by the sink "anti-loudspeaker". The finite wavetrain allows us to see the effect of pseudomomentum changes at $O(\epsilon)$. This set-up is described in the somewhat busy Fig. 5.7.

The figure shows the loudspeakers and the vortex at a distance D from the steady wavetrain. The circumferential basic velocity with magnitude $\tilde{U}$ is indicated by the dashed line. The loudspeakers are slightly angled because of the mean flow

---

[13] The azimuthal symmetry of this flow induces a further ray invariance, namely $lx - ky$. However, this "angular momentum" invariant is not needed here.

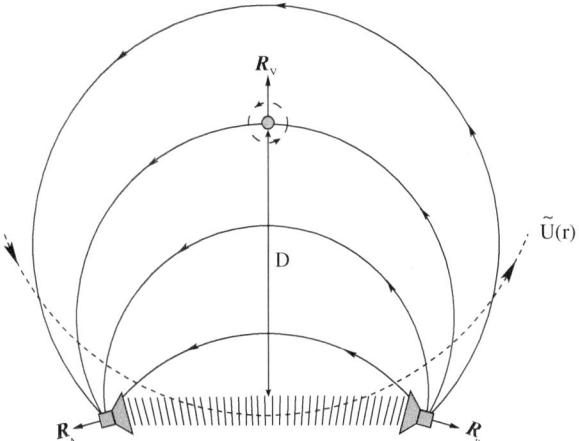

Fig. 5.7 A steady wavetrain travels *left* to *right* from the wavemaker A to the wave absorber B. The vortex refracts the wavetrain and there is a net recoil force $\boldsymbol{R}_A + \boldsymbol{R}_B$ in the y-direction on the loudspeakers. Concomitant is an effective remote recoil force $\boldsymbol{R}_V$ felt by the vortex such that $\boldsymbol{R}_A + \boldsymbol{R}_B + \boldsymbol{R}_V = 0$. The recoil is mediated by a leftward material displacement of the vortex at $O(a^2)$ due to the Bretherton return flow

refraction and the wave crests are counter-rotating relative to the vortex. One can see by inspection that the y-component of the intrinsic group velocity at the wavemaker A on the left cancels the y-component of the basic velocity there and vice versa at the wave absorber B on the right.

Now, at the irrotational loudspeakers we have

$$\overline{\boldsymbol{F}}^L = \mathcal{F} \quad \text{and} \quad \boldsymbol{R}_{A,B} = -\int_{A,B} \overline{\boldsymbol{F}}^L \, dx dy \tag{5.86}$$

are the respective recoil forces exerted on the wavemaker and wave absorber (we use $H = 1$ for simplicity). There is an equal-and-opposite push in the x-direction and a net recoil in the negative y-direction due to the refraction, i.e.

$$\boldsymbol{R}_A + \boldsymbol{R}_B + \boldsymbol{R}_V = 0 \quad \text{where} \quad \boldsymbol{R}_V = (0, R_V) \quad \text{with} \quad R_V > 0, \tag{5.87}$$

say. The net recoil $-\boldsymbol{R}_V$ equals the net pseudomomentum generation per unit time due to the refraction.

How does the mean flow impulse change? The impulse plus pseudomomentum conservation law (5.56) for a steady wavetrain yields

$$\frac{d(\mathcal{I} + \mathcal{P})}{dt} = \frac{d\mathcal{I}}{dt} = \int \overline{\boldsymbol{F}}^L \, dx dy = -(\boldsymbol{R}_A + \boldsymbol{R}_B) = \boldsymbol{R}_V. \tag{5.88}$$

This shows that the mean flow impulse should change in order to compensate for the net recoil in the y-direction exerted on the loudspeakers. The total impulse due to a single vortex with nonzero net circulation depends on the origin of the coordinate system, but *changes* in the impulse due to movement of the vortex are coordinate-independent. In particular, the mean flow impulse is

$$\mathcal{I} = \int (y, -x)\bar{q}^L \, dxdy = (Y, -X)\Gamma \qquad (5.89)$$

if (X, Y) are the coordinates of the vortex centroid. Therefore (5.88) implies

$$\frac{dY}{dt} = 0 \quad \text{and} \quad \frac{dX}{dt} = -\frac{R_V}{\Gamma}. \qquad (5.90)$$

This is a surprising result because it means that the vortex must move to the left in Fig. 5.7. Where does the velocity field come from that achieves this material displacement?

The answer comes from the wave–mean response at $O(a^2)$ to the presence of the finite wavetrain. In fact, to compute this response at leading order it is not necessary to include the vortex. Without the vortex there is no tilt in the loudspeakers and the wave crests do not rotate. The Lagrangian mean flow in the presence of the irrotational steady wavetrain is then determined from

$$\nabla \cdot \bar{u}^L = 0 \quad \text{and} \quad \bar{q}^L = 0 \quad \Rightarrow \quad \nabla \times \bar{u}^L = \nabla \times \mathbf{p}. \qquad (5.91)$$

These equations express that $\bar{u}^L$ is the least-squares projection of $\mathbf{p}$ onto the space of non-divergent vector fields. The pseudomomentum curl is positive above the wavetrain and negative below, so $\bar{u}^L$ is the incompressible flow described by a vortex couple with precisely this curl. Away from the wavetrain $\bar{u}^L = \bar{u}$ and the flow resembles the standard dipole flow familiar from elementary fluid mechanics. The streamlines of this dipole flow are indicated by the solid lines in Fig. 5.7; we suggested the name "Bretherton flow" for this characteristic mean flow response to a wavepacket, because its description goes back to [8].

Crucially, the Bretherton flow points backward, i.e. in the negative x-direction, at the vortex location above the wavetrain. Therefore the Bretherton flow does indeed push the vortex to the left and it has been checked in [16] that it does so with precisely the right magnitude to be consistent with (5.90).

We called this action-at-a-distance of the wavetrain on the vortex "remote recoil" in order to stress the non-local nature of this wave–vortex interaction. After all, the waves and the vortex do not overlap in physical space. The term "recoil" is also apt because the movement of the vortex is consistent with the action of a compact body force on the vortex with net integral equal to R_V. Such a force would be relevant in a parametrization problem in which the small-scale wavetrain is not explicitly resolved but modelled. This force will produce positive vorticity to the left of the vortex and negative vorticity to its right, which would lead to a movement of the vortex centroid to the left as required. In this case the vorticity moves although the fluid particles do not.

Finally, it can be shown that the remote recoil idea remains valid at $O(\epsilon^2)$ if the loudspeakers recede to infinity and the pseudomomentum generation is due to the weak $O(\epsilon^2)$ net scattering of the waves into the lee of the vortex.

So whilst the set-up in Fig. 5.7 is certainly very special it is not artificial; the recoil is real.

5.5.4 Wave Capture of Internal Gravity Waves

The analogy between tracer advection and wave refraction described in Sect. 5.5.1 suggests a further interesting possibility for wave–vortex interactions: the unbounded exponential growth in time of $\boldsymbol{k}$ and therefore the unbounded growth of $\mathcal{P}$ which is implied by the conservation of wave action and the definition of pseudomomentum for a wavetrain. Such exponential growth is the likely outcome in the long run in the passive tracer case, although a proper description of this process also requires statistical tools. A convenient reference point in the literature aimed at related atmospheric applications is [23].

However, unbounded growth of $\boldsymbol{k}$ is not possible in the shallow-water wave system, at least not in the case of steady and sub-critical $\boldsymbol{U}(\boldsymbol{x})$. This is because then the ray invariance of ω implies a bound on κ:

$$\omega = \omega_0 = \boldsymbol{U} \cdot \boldsymbol{k} + c\kappa \quad \Rightarrow \quad \kappa \le \frac{\omega_0/c}{1 - ||\boldsymbol{U}/c||_\infty}. \tag{5.92}$$

This is finite if the maximal Froude/Mach number $||\boldsymbol{U}/c||_\infty$ is less than unity so for sub-critical flows there can be no unbounded wavenumber growth. However, this is not the case for other types of waves.

A good example of geophysical interest are internal gravity waves in the three-dimensional Boussinesq system, which is described by

$$\nabla \cdot \boldsymbol{u} = 0, \quad \frac{Db}{Dt} + N^2 w = 0, \quad \text{and} \quad \frac{D\boldsymbol{u}}{Dt} + \nabla P = b\widehat{z}. \tag{5.93}$$

Here $\boldsymbol{u} = (u, v, w)$ is the velocity, b is the buoyancy, N is the buoyancy frequency, P is the pressure divided by the constant reference density, and $\widehat{z}$ is the unit vector in the vertical. All fields depend on $\boldsymbol{x} = (x, y, z)$ and t. The buoyancy equation can also be written as

$$\frac{D\Theta}{Dt} = 0 \quad \text{with} \quad \Theta = b + N^2 z, \tag{5.94}$$

which shows that Θ is a material invariant marking the stratification surfaces. Physically, Θ corresponds to either potential temperature or density in the atmosphere or ocean. Note that (5.94) implies the exact $\overline{D}^L \overline{\Theta}^L = 0$.

Linear internal waves are dispersive transversal waves (due to $\nabla \cdot \boldsymbol{u} = 0$) with dispersion relation

$$\omega = U \cdot k \pm N \sqrt{\frac{k^2 + l^2}{k^2 + l^2 + m^2}}, \qquad (5.95)$$

where $k = (k, l, m)$. The intrinsic part of this dispersion relation is zeroth-order homogeneous in the wavenumber components, so multiplication of k by any nonzero constant does not change $\hat{\omega}$. Consequently, there is no a priori bound on wavenumber growth in this case. There is no leading-order Stokes drift for slowly varying wavetrains in this system but the pseudomomentum is still given by the generic formula

$$\mathbf{p} = \frac{k}{\hat{\omega}} \bar{E} \quad \text{where} \quad \bar{E} = \frac{1}{2} \left(\overline{u'^2} + \overline{v'^2} + \overline{w'^2} + \frac{\overline{b'^2}}{N^2} \right) = \overline{|u'|^2} = \frac{\overline{b'^2}}{N^2}. \qquad (5.96)$$

Now, U could have three nonzero components but for atmosphere–ocean applications a useful restriction is to consider $U = (U, V, 0)$ with $U_x + V_y = 0$, which models the quasi-horizontal layerwise flow familiar from quasi-geostrophic dynamics (e.g. [23]). (We omitted Coriolis forces in (5.93), but they are easily added there and in the remainder of the theory; cf. [17].)

This restriction implies that the refraction problems for the horizontal wavenumbers $k_H = (k, l, 0)$ and for the vertical wavenumber m decouple, i.e. we find that

$$\frac{d k_H}{dt} = -(\nabla_H U) \cdot k_H \quad \text{and} \quad \frac{dm}{dt} = -k U_z - l V_z, \qquad (5.97)$$

where $\nabla_H = (\partial_x, \partial_y, 0)$. The horizontal part evolves completely analogous to the two-dimensional considerations in Sect. 5.5.1. If circumstances conspire (i.e. if the basic flow is varying slowly enough along a group velocity ray) then k_H will be turned onto a growing eigenvector of (5.97) and exponential growth in k_H and m will follow. This leads to an interesting phenomenon ([4, 17]) which is due to the fact that the magnitude of the intrinsic group velocity of internal gravity waves at fixed intrinsic frequency is proportional to the wavelength of the wave. This follows immediately from the zeroth-order homogeneity of $\hat{\omega}$ as a function of k, because

$$\hat{c}_g = \frac{\partial \hat{\omega}}{\partial k}, \qquad (5.98)$$

and therefore $\hat{c}_g$ is homogeneous in k of order minus one. Explicitly, we have

$$|\hat{c}_g| = \frac{1}{\kappa} \sqrt{\frac{N^2 - \hat{\omega}^2}{\hat{\omega}^2}}. \qquad (5.99)$$

This suggests that exponential growth of k goes together with $\hat{\omega}$ approaching a constant and $|\hat{c}_g| \to 0$ exponentially. In other words, the wavepacket gets "glued" into the basic flow [4], because its group velocity converges to the basic flow velocity. By definition, this strengthens the analogy between passive advection and wave refraction, which then leads to more stretching of $\boldsymbol{k}$ and to even more reduced $|\hat{\boldsymbol{c}}_g|$, reinforcing the cycle.[14]

This process and the attendant wave–vortex interactions were studied under the name "wave capture" in [17]. The key question is: How does the mean flow react to the exponentially growing amount of pseudomomentum $\mathcal{P}$ that is contained in a wavepacket? The answer to this question follows reasonably easily once we have written down the impulse plus pseudomomentum conservation law for three-dimensional stratified flow.

5.5.5 Impulse Plus Pseudomomentum for Stratified Flow

This is discussed in detail in [17], so we only summarize the result. Basically, it is possible to write down a useful impulse for the *horizontal* mean flow in the Boussinesq system provided the mean stratification surfaces remain almost flat in the chosen coordinate system. Specifically, we assume that

$$\nabla_H \cdot \overline{\boldsymbol{u}}_H^L = 0 \quad \text{and} \quad \overline{w}^L = 0, \tag{5.100}$$

and also that the mean stratification surfaces $\overline{\Theta}^L = \text{constant}$ are horizontal planes to sufficient approximation. There is an exact GLM PV law

$$\tilde{\rho}\overline{q}^L = \nabla\overline{\Theta}^L \cdot \nabla \times (\overline{\boldsymbol{u}}^L - \mathbf{p}) \quad \Rightarrow \quad \overline{D}^L \overline{q}^L = 0 \tag{5.101}$$

if $\overline{D}^L \tilde{\rho} + \tilde{\rho}\nabla \cdot \overline{\boldsymbol{u}}^L = 0$, but with the above assumption we have the simpler

$$\overline{q}^L = \hat{z} \cdot \nabla \times (\overline{\boldsymbol{u}}^L - \mathbf{p}) \equiv \nabla_H \times (\overline{\boldsymbol{u}}_H^L - \mathbf{p}_H). \tag{5.102}$$

We can now define the total horizontal mean flow impulse and pseudomomentum by

$$\mathcal{I}_H = \int (y, -x, 0)\overline{q}^L \, dxdydz \quad \text{and} \quad \mathcal{P}_H = \int \mathbf{p}_H \, dxdydz \tag{5.103}$$

and we then find the conservation law

$$\frac{d(\mathcal{I}_H + \mathcal{P}_H)}{dt} = \int \overline{\boldsymbol{F}}_H^L \, dxdydz. \tag{5.104}$$

---

[14] The slowdown of the wavepacket is reminiscent of the well-known shear-induced critical layers, which inhibit vertical propagation past a certain critical line. Still, the details are quite different, e.g. here the wavenumber grows exponentially in time whereas in the classical critical layer scenario it grows linearly in time.

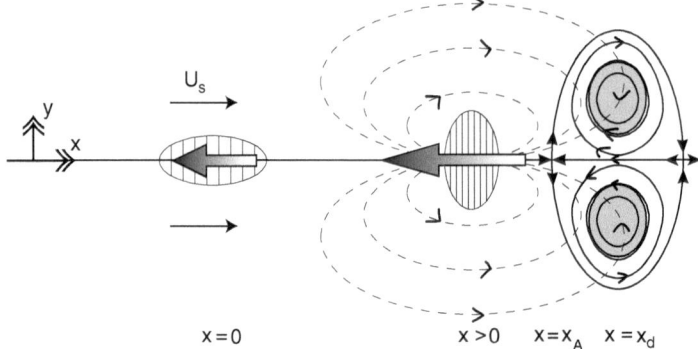

Fig. 5.8 A wavepacket indicated by the wave crests and arrow for the net pseudomomentum is
squeezed by the straining flow due to a vortex couple on the *right*. The vortex couple travels a little
faster than the wavepacket, so the wavepacket slides toward the stagnation point in front of the
couple, its x-extent decreases, its y-extent increases, and so does its total pseudomomentum. The
pseudomomentum increase is compensated by a decrease in the vortex couple impulse caused by
the Bretherton flow of the wavepacket, which is indicated by the *dashed lines*

As before, both $\mathcal{I}_H$ and $\mathcal{P}_H$ vary individually due to refraction and momentum-
conserving dissipation, but their sum remains constant unless the flow is forced
externally.

This makes obvious that during wave capture any exponential growth of $\mathcal{P}_H$
must be compensated by an exponential decay of $\mathcal{I}_H$. Because the value of $\overline{q}^L$ on
mean trajectories cannot change, this must be achieved via material displacements
of the PV structure, just as in the remote recoil situation in shallow water.

As an example we consider the refraction of a wavepacket by a vortex couple
as in Fig. 5.8, which shows a horizontal cross-section of the flow [17]. The area-
preserving straining flow due to the vortex dipole increases the pseudomomentum
of the wavepacket, because it compresses the wavepacket in the x-direction whilst
stretching it in the y-direction. At the same time, the Bretherton flow induced by
the finite wavepacket pushes the vortex dipole closer together, which reduces the
impulse of the couple and this is how (5.104) is satisfied.

5.5.6 Local Mean Flow Amplitude at the Wavepacket

The previous considerations made clear that the exponential surge in packet-integr-
ated pseudomomentum is compensated by the loss of impulse of the vortex cou-
ple far away. Still, there is a lingering concern about the local structure of $\overline{u}^L$ at
the wavepacket. For instance, the exact GLM relation (5.16) for periodic zonally
symmetric flows suggests that $\overline{u}^L$ at the core of the wavepacket might make a large
amplitude excursion because it might follow the local pseudomomentum p_1, which
is growing exponentially in time. This is an important consideration, because a large
$\overline{u}^L$ might induce wave breaking or other effects.

We can study this problem easily in a simple two-dimensional set-up, brushing aside concerns that our two-dimensional theory may be misleading for the three-dimensional stratified case. In particular, we look at a wavepacket centred at the origin of an (x, y) coordinate system such that at $t = 0$ the pseudomomentum is $\mathbf{p} = (1, 0) f(x, y)$ for some envelope function f that is proportional to the wave action density. This is the same wavepacket set-up as in Sect. 5.3.3. At all times the local Lagrangian mean flow at $O(a^2)$ induced by the wavepacket is the Bretherton flow, which by $\overline{q}^L = 0$ is the solution of

$$\overline{u}^L_x + \overline{v}^L_y = 0 \quad \text{and} \quad \overline{v}^L_x - \overline{u}^L_y = \nabla \times \mathbf{p} = -f_y(x, y). \tag{5.105}$$

We imagine that the wavepacket is exposed to a pure straining basic flow $\mathbf{U} = (-x, +y)$, which squeezes the wavepacket in x and stretches it in y. We ignore intrinsic wave propagation relative to $\mathbf{U}$, which implies that the wave action density f is advected by $\mathbf{U}$, i.e. $D_t f = 0$. We then obtain the refracted pseudomomentum as

$$\mathbf{p} = (\alpha, 0) f(\alpha x, y/\alpha) \quad \text{and} \quad \nabla \times \mathbf{p} = -f_y(\alpha x, y/\alpha). \tag{5.106}$$

Here $\alpha = \exp(t) \geq 1$ is the scale factor at time $t \geq 0$ and (5.106) shows that p_1 grows exponentially whilst $\nabla \times \mathbf{p}$ does not; in fact $\nabla \times \mathbf{p}$ is materially advected by $\mathbf{U}$, just as the wave action density f and unlike the pseudomomentum density $\mathbf{p}$. This is a consequence of the stretching in the transverse y-direction, which diminishes the curl because it makes the x-pseudomomentum vary more slowly in y. Thus whilst there is an exponential surge in p_1 there is none in $\nabla \times \mathbf{p}$.

In an unbounded domain we can go one step further and explicitly compute $\overline{u}^L$ at the core of the wavepacket, say. We use Fourier transforms defined by

$$\text{FT}\{f\}(k, l) = \int e^{-i[kx+ly]} f(x, y) \, dx dy \tag{5.107}$$

and

$$f(x, y) = \frac{1}{4\pi^2} \int e^{+i[kx+ly]} \text{FT}\{f\}(k, l) \, dk dl. \tag{5.108}$$

The transforms of $\overline{u}^L$ and of p_1 are related by

$$\text{FT}\{\overline{u}^L\}(k, l) = \frac{l^2}{k^2 + l^2} \text{FT}\{\mathsf{p}_1\}(k, l). \tag{5.109}$$

This follows from $\mathbf{p} = (\mathsf{p}_1, 0)$ and the intermediate introduction of a stream function ψ such that $(\overline{u}^L, \overline{v}^L) = (-\psi_y, +\psi_x)$ and therefore $\nabla^2 \psi = -\mathsf{p}_{1y}$. The scale-insensitive pre-factor varies between zero and one and quantifies the projection onto non-divergent vector fields in the present case. This relation by itself does not rule out exponential growth of $\bar{u}^L$ in some proportion to the exponential growth of p_1. We need to look at the spectral support of p_1 as the refraction proceeds.

We denote the initial p_1 for $\alpha = 1$ by p_1^1 and then the pseudomomentum for other values of α is $\mathsf{p}_1^\alpha(x, y) = \alpha \mathsf{p}_1^1(\alpha x, y/\alpha)$. The transform is found to be

$$\mathrm{FT}\{\mathsf{p}_1^\alpha\}(k, l) = \alpha \mathrm{FT}\{\mathsf{p}_1^1\}(k/\alpha, \alpha l). \qquad (5.110)$$

This shows that with increasing α the spectral support shifts towards higher values of k and lower values of l. The value of $\bar{u}^L$ at the wavepacket core $x = y = 0$ is the total integral of (5.109) over the spectral plane, which using (5.110) can be written as

$$\bar{u}^L(0, 0) = \frac{1}{4\pi^2} \int \frac{l^2}{k^2 + l^2} \mathrm{FT}\{\mathsf{p}_1^\alpha\}(k, l)\, dk dl$$

$$= \frac{\alpha}{4\pi^2} \int \frac{l^2}{\alpha^4 k^2 + l^2} \mathrm{FT}\{\mathsf{p}_1^1\}(k, l)\, dk dl \qquad (5.111)$$

after renaming the dummy integration variables. This is as far as we can go without making further assumptions about the shape of the initial wavepacket.

For instance, if the wavepacket is circularly symmetric initially, then p_1^1 depends only on the radius $r = \sqrt{x^2 + y^2}$ and $\mathrm{FT}\{\mathsf{p}_1^1\}$ depends only on the spectral radius $\kappa = \sqrt{k^2 + l^2}$. In this case (5.111) can be explicitly evaluated by integrating over the angle in spectral space and yields the simple formula

$$\bar{u}^L(0, 0) = \frac{\alpha}{\alpha^2 + 1} \mathsf{p}_1^1(0, 0) = \frac{1}{\alpha^2 + 1} \mathsf{p}_1^\alpha(0, 0). \qquad (5.112)$$

The pre-factor in the first expression has maximum value $1/2$ at $\alpha = 1$, which implies that the maximal Lagrangian mean velocity at the wavepacket core is the initial velocity, when the wavepacket is circular. At this initial time $\bar{u}^L = 0.5\mathsf{p}_1$ at the core and thereafter $\bar{u}^L$ decays; there is no growth at all.

So this proves that there is no surge of local mean velocity even though there is a surge of local pseudomomentum. This simple example serves as a useful illustration of how misleading zonally symmetric wave–mean interaction theory can be when we try to understand more general wave–vortex interactions.

Finally, how about a wavepacket that is not circularly symmetric at $t = 0$? The worst case scenario is an initial wavepacket that is long in x and narrow in y; this corresponds to values of α near zero and the second expression in (5.112) then shows that the mean velocity at the core is almost equal to the pseudomomentum. This scenario recovers the predictions of zonally symmetric theory.

The subsequent squeezing in x now amplifies the pseudomomentum and this leads to a transient growth of $\bar{u}^L$ in proportion, at least whilst the wavepacket still has approximately the initial aspect ratio. However, eventually the aspect ratio reverses and the wavepacket becomes short in x and wide in y; this corresponds to α much larger than unity. Eventually α becomes large and $\overline{u}^L$ decays as $1/\alpha = \exp(-t)$.

5.5.7 Wave–Vortex Duality and Dissipation

We take another look at the similarity between a wavepacket and a vortex couple in an essentially two-dimensional situation (see Fig. 5.9). The Bretherton flow belonging to the wavepacket is described by (5.105). In the three-dimensional Boussinesq system the Bretherton flow is observed on any stratification surface currently intersected by a compact wavepacket [8]. The physical reason for this different behaviour is the infinite adjustment speed related to pressure forces in the incompressible Boussinesq system; such infinitely fast action-at-a-distance is not available in the shallow water system. We will look at the three-dimensional stratified case.

Now, the upshot of this is that a propagating wavepacket gives rise to a mean flow that instantaneously looks identical to that of a vortex couple with vertical vorticity equal to $\nabla_H \times \mathbf{p}_H$. Of course, this peculiar vortex couple attached to the wavepacket moves with the group velocity, not with the nonlinear material velocity. Importantly, refraction can change the wavepacket's pseudomomentum curl in a manner that is again identical to that of a vortex couple, a situation that is particularly clear during wave capture. For instance, in Fig. 5.8 the straining of the captured wavepacket leads to the material advection of pseudomomentum curl, just as in a vortex couple. If the wavepacket were to be replaced by that vortex couple, then we would recognize that Fig. 5.8 displays the early stage of the classical vortex-ring leap-frogging dynamics, with two-dimensional vortex couples replacing the three-dimensional vortex rings of the classical example. This suggests a "wave–vortex

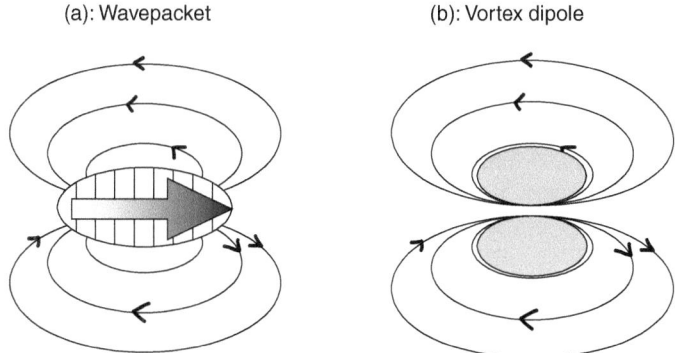

Fig. 5.9 Wave–vortex duality. *Left:* wavepacket together with streamlines indicating the Bretherton flow; the *arrow* indicates the net pseudomomentum. *Right:* a vortex couple with the same return flow; the *shaded areas* indicate nonzero PV values with opposite signs

duality", because the wavepacket acts and interacts with the remaining flow *as if* it were a vortex couple.

Moreover, if we allow the wavepacket to dissipate, then the wavepacket on the left in Fig. 5.9 would simply turn into the dual vortex couple on the right in terms of the structural changes in $\overline{q}^L$ that occur during dissipation. However, there would be no mean flow acceleration during the dissipation, for the same reasons that were discussed in Sect. 5.3.4. This leads to an intriguing consideration: if a three-dimensional wavepacket has been captured by the mean flow (i.e. its intrinsic group velocity has become negligible), then whether or not the wavepacket dissipates has no effect on the mean flow [17].

These considerations lead to a view of wave capture as a peculiar form of dissipation: the loss of intrinsic group velocity is equivalent, as far as wave–vortex interactions are concerned, to the loss of the wavepacket altogether.

5.6 Concluding Comments

All the theoretical arguments and examples presented in this chapter served to illustrate the interplay between wave dynamics and PV dynamics during strong wave–vortex interactions. Only highly idealized flow situations were considered in order to stress the fundamental aspects of the fluid dynamics whilst reducing clutter in the equations. For instance, Coriolis forces were neglected throughout, but they can be incorporated both in GLM theory and in the other theoretical developments; this has been done in the quoted references.

The main difference between the results presented here and those available in the textbooks on geophysical fluid dynamics [e.g. 39, 42] is that we have not used the twin assumptions of zonal periodicity and zonal mean flow symmetry, which are the starting points of most accounts of wave–mean interaction theory in the literature. As is well known, these assumptions work well for zonal-mean atmospheric flows, but they do not work for most oceanic flows (away from the Antarctic circumpolar current, say), which are typically hemmed-in by the continents and therefore are not periodic. To understand local wave–mean interactions in such geometries requires different tools.

In practice, even when zonal mean theory is applicable it might not use the best definition of a mean flow. For instance, in general circulation models (GCMs) it is natural to think of the resolved large-scale flow as the mean flow and of the unresolved sub-grid-scale motions as the disturbances. This suggests local averaging over grid boxes rather than global averaging over latitude circles. This has an impact on the parametrization of unresolved wave motions in such GCMs, which are typically applied to each grid column in isolation even though their theoretical underpinning is typically based on zonally symmetric mean flows. For example, in [22] the global angular momentum transport due to atmospheric gravity waves in a model that allows for three-dimensional refraction effects is compared against a traditional parametrization based on zonally symmetric mean flows.

From a fundamental viewpoint, all wave–mean interaction theories seek to simplify the mean pressure forces in the equations. The reason is that the pressure is difficult to control both physically and mathematically, because it reacts rapidly and at large distances to changes and excitations of the flow, both wavelike and vortical. In zonal-mean theory for periodic flows the net zonal pressure force drops out of the zonal momentum equations, but this does not work in the local version of the problem. On the other hand, Kelvin's circulation theorem and potential vorticity dynamics are independent of pressure forces from the outset. Thus, quite naturally, whilst zonal-mean theory is based on zonal momentum, the local wave–mean interaction theory presented here is based on potential vorticity.

Perhaps the single most important message from this chapter is the role played by the pseudomomentum vector in the mean circulation theorem (5.15). All subsequent results flow from this theorem, which shows why pseudomomentum is so important in wave–mean interaction theory. This contrasts with the primary stress that is often placed on the integral conservation of pseudomomentum in the presence of translational mean flow symmetries.

We now know that pseudomomentum plays a crucial role in wave–mean interaction theory whether or not specific components of it are conserved.

Acknowledgments It is a pleasure to thank the organizers of the Alpine Summer School 2006 in Aosta (Italy) for their kind invitation to deliver the lectures on which this chapter is based. This research is supported by the grants OCE-0324934 and DMS-0604519 of the National Science Foundation of the USA. I would also like to acknowledge the kind hospitality of the Zuse Zentrum at the Freie Universität Berlin (Germany) during my 2007 sabbatical year when this chapter was written.

References

1. Andrews, D.G., Holton, J.R., Leovy, C.B.: Middle Atmosphere Dynamics. Academic, New York (1987).
2. Andrews, D.G., McIntyre, M.E.: An exact theory of nonlinear waves on a Lagrangian-mean flow. J. Fluid Mech. **89**, 609–646 (1978).
3. Andrews, D.G., McIntyre, M.E.: On wave-action and its relatives. J. Fluid Mech. **89**, 647–664 (1978).
4. Badulin, S.I., Shrira, V.I.: On the irreversibility of internal waves dynamics due to wave trapping by mean flow inhomogeneities. Part 1. Local analysis. J. Fluid Mech. **251**, 21–53 (1993).
5. Barreiro, A.K., Bühler, O.: Longshore current dislocation on barred beaches. J. Geophys. Res. Oceans **113**, C12004 (2008).
6. Batchelor, G.K.: An Introduction to Fluid Dynamics. Cambridge University Press, Cambridge (1967).
7. Bouchut, F., Le Sommer, J., Zeitlin, V.: Breaking of balanced and unbalanced equatorial waves. Chaos **15**, 3503 (2005).
8. Bretherton, F.P.: On the mean motion induced by internal gravity waves. J. Fluid Mech. **36**, 785–803 (1969).
9. Brocchini, M., Kennedy, A., Soldini, L., Mancinelli, A.: Topographically controlled, breakingwave-induced macrovortices. Part 1. Widely separated breakwaters. J. Fluid Mech. **507**, 289–307 (2004).
10. Bühler, O.: A shallow-water model that prevents nonlinear steepening of gravity waves. J. Atmos. Sci. **55**, 2884–2891 (1998).

11. Bühler, O.: On the vorticity transport due to dissipating or breaking waves in shallow-water flow. J. Fluid Mech. **407**, 235–263 (2000).
12. Bühler, O.: Impulsive fluid forcing and water strider locomotion. J. Fluid Mech. **573**, 211–236 (2007).
13. Bühler, O.: Waves and Mean Flows. Cambridge University Press, Cambridge (2008).
14. Bühler, O., Jacobson, T.E.: Wave-driven currents and vortex dynamics on barred beaches. J. Fluid Mech. **449**, 313–339 (2001).
15. Bühler, O., McIntyre, M.E.: On non-dissipative wave–mean interactions in the atmosphere or oceans. J. Fluid Mech. **354**, 301–343 (1998).
16. Bühler, O., McIntyre, M.E.: Remote recoil: a new wave–mean interaction effect. J. Fluid Mech. **492**, 207–230 (2003).
17. Bühler, O., McIntyre, M.E.: Wave capture and wave–vortex duality. J. Fluid Mech. **534**, 67–95 (2005).
18. Church, J.C., Thornton, E.B.: Effects of breaking wave induced turbulence within a longshore current model. Coast. Eng. **20**, 1–28 (1993).
19. Drucker, E.G., Lauder, G.V.: Locomotor forces on a swimming fish: three-dimensional vortex wake dynamics quantified using digital particle image velocimetry. J. Exp. Biol. **202**, 2393–2412 (1999).
20. Dysthe, K.B.: Refraction of gravity waves by weak current gradients. J. Fluid Mech. **442**, 157–159 (2001).
21. Grianik, N., Held, I.M., Smith, K.S., Vallis, G.K.: The effects of quadratic drag on the inverse cascade of two-dimensional turbulence. Phys. Fluids **16**, 73–78 (2004).
22. Hasha, A.E., Bühler, O., Scinocca, J.F.: Gravity-wave refraction by three-dimensionally varying winds and the global transport of angular momentum. J. Atmos. Sci. **65**, 2892–2906 (2008).
23. Haynes, P.H., Anglade, J.: The vertical-scale cascade of atmospheric tracers due to large-scale differential advection. J. Atmos. Sci. **54**, 1121–1136 (1997).
24. Haynes, P.H., McIntyre, M.E.: On the conservation and impermeability theorems for potential vorticity. J. Atmos. Sci. **47**, 2021–2031 (1990).
25. Hinds, A.K., Johnson, E.R., McDonald, N.R.: Vortex scattering by step topography. J. Fluid Mech. **571**, 495–505 (2007).
26. Johnson, E.R., Hinds, A.K., McDonald, N.R.: Steadily translating vortices near step topography. Phys. Fluids **17**, 6601 (2005).
27. Jones, W.L.: Ray tracing for internal gravity waves. J. Geophys. Res. **74**, 2028–2033 (1969).
28. Kennedy, A.B., Brocchini, M., Soldini, L., Gutierrez, E.: Topographically controlled, breakingwave-induced macrovortices. Part 2. Changing geometries. J. Fluid Mech. **559**, 57–80 (2006).
29. Kuo, A., Polvani, L.M.: Wave-vortex interactions in rotating shallow water. Part I. One space dimension. J. Fluid Mech. **394**, 1–27 (1999).
30. Lamb, H.: Hydrodynamics, 6th edn. Cambridge University Press, Cambridge (1932).
31. Landau, L.D., Lifshitz, E.M.: Mechanics, 3rd English edn. Butterworth–Heinemann, Oxford (1982).
32. Longuet-Higgins, M.S.: Longshore currents generated by obliquely incident sea waves 1. J. Geophys. Res. **75**, 6778–6789 (1970).
33. Longuet-Higgins, M.S.: Longshore currents generated by obliquely incident sea waves 2. J. Geophys. Res. **75**, 6790–6801 (1970).
34. McIntyre, M.E.: Balanced atmosphere-ocean dynamics, generalized lighthill radiation, and the slow quasi-manifold. Theor. Comput. Fluid Dyn. **10**, 263–276 (1998).
35. McIntyre, M.E.: On global-scale atmospheric circulations. In: Batchelor, G.K., Moffatt, H.K., Worster, M.G. (eds.) Perspectives in Fluid Dynamics: A Collective Introduction to Current Research, 631 pp. Cambridge University Press, Cambridge (2003).
36. McIntyre, M.E., Norton, W.A.: Dissipative wave–mean interactions and the transport of vorticity or potential vorticity. J. Fluid Mech. **212**, 403–435 (1990).

37. Peregrine, D.H.: Surf zone currents. Theor. Comput. Fluid Dyn. **10**, 295–310 (1998).
38. Peregrine, D.H.: Large-scale vorticity generation by breakers in shallow and deep water. Eur. J. Mech. B/Fluids **18**, 403–408 (1999).
39. Salmon, R.: Lectures on Geophysical Fluid Dynamics. Oxford University Press, Oxford (1998).
40. Theodorsen, T.: Impulse and momentum in an infinite fluid. In: Theodore Von Karman Anniversary Volume, pp. 49–57. Caltech, Pasadena (1941).
41. Vadas, S.L., Fritts, D.C.: Gravity wave radiation and mean responses to local body forces in the atmosphere. J. Atmos. Sci. **58**, 2249–2279 (2001).
42. Vallis, G.K.: Atmospheric and Oceanic Fluid Dynamics: Fundamentals and Large-Scale Circulation. Cambridge University Press, Cambridge (2006).
43. Wunsch, C., Ferrari, R.: Vertical mixing, energy, and the general circulation of the oceans. Annu. Rev. Fluid Mech. **36**, 281–314 (2004).
44. Zhu, X., Holton, J.: Mean fields induced by local gravity-wave forcing in the middle atmosphere. J. Atmos. Sci. **44**, 620–630 (1987).

Index

Flór, J.-B. (ed.): *Index*. Lect. Notes Phys. **805**, 189–192 (2010)
DOI 10.1007/978-3-642-11587-5   © Springer-Verlag Berlin Heidelberg 2010